Principles of
Tropical Agronomy

S.N. Azam-Ali

Tropical Crops Research Unit
University of Nottingham
School of Biosciences
Sutton Bonington
UK

and

G.R. Squire

Scottish Crop Research Institute
Invergowrie
Dundee
UK

CABI *Publishing*

CABI *Publishing* **is a division of CAB** *International*

CABI Publishing
CAB International
Wallingford
Oxon OX10 8DE
UK

Tel: +44 (0)1491 832111
Fax: +44 (0)1491 833508
Email: cabi@cabi.org
Web site www.cabi.org

CABI Publishing
10 E 40th Street
Suite 3203
New York, NY 10016
USA

Tel: +1 212 481 7018
Fax: +1 212 686 7993
Email: cabi-nao@cabi.org

A catalogue record for this book is available from the British Library, London,
UK.

Library of Congress Cataloging-in-Publication Data
Azam-Ali, S. N. (Sayed N.)
    Principles of tropical agronomy/S. N. Azam-Ali and G. R. Squire.
        p. cm.
    Includes bibliographical references.
    ISBN 0-85199-136-X (pbk.)
        1. Agriculture–Tropics. I. Squire, G. R. (Geoffrey R.) II. Title.
    S604.37. A94 2001

                                                                2001037838

ISBN 0 85199 136 X

Typeset in Helvetica and Melior by Wyvern 21 Ltd, Bristol
Printed and bound in the UK by Biddles Ltd, Guildford and King's Lynn

Contents

Preface

This book is concerned with the principles by which crop management can influence how plants use environmental resources. It deals with the biological processes that operate in the links between management, the physical environment and agricultural vegetation. It also seeks to identify general principles that operate across tropical species, cropping systems and environ- ments. The purpose of the book is to pro- vide students, researchers, advisers and managers with a biologically based account that is nevertheless accessible to those with little knowledge of plant biology and en- vironmental physics. It is hoped that this framework of analysis will enable more informed decisions to be made both about applied research and about management.

Acknowledgements

A number of people have provided constructive suggestions and valuable help during the preparation of various chapters in this volume and in the presentation of the text, references and diagrams. In particular, we wish to thank David Hodson, Festo Massawe, Rutger Cornelissen, John Foulkes and Sheila Northover.

1

Contexts

Agriculture in the tropics is complex and diverse, encompassing plants and soils, climate and people, trade and survival. It can be very diverse biologically, as in mixed gardens; elsewhere, one species can dominate vast areas of land. In dry, marginal areas, crops may grow for only a few weeks each year when there is sufficient water and, often, their yields are barely adequate to sustain life. In humid, fertile regions, plants yield more-or-less continuously, giving products for exports that dominate the economy.

Despite this diversity, common elements allow the terms 'tropical agriculture' and 'tropical agronomy'. The geographical location between the Tropics of Cancer and Capricorn is an approximation, but the warmth and the bright sunlight between and around these limits are what most people associate with the word 'tropical'. To the agriculturalist and ecologist, the soils and plants provide further similarities between continents and countries. Many tropical soils are potentially unstable, low in fertility, easily eroded and shallow. The cereals, legumes, roots, fibres and fruits that are normally recognized as tropical crops are often grown by similar forms of husbandry in Africa, Asia, the Pacific and the Americas. Of all factors, temperature most strongly limits the extent of these species. Most develop and grow between about 10 and 40°C and yield economically at mean monthly temperatures between 15 and 30°C. Very few have any chance of surviving to yield in the cold or frosty weather of temperate lands in winter.

However, there are important exceptions to these generalities. Several widespread subsistence crops such as potato, and certain cereals including wheat and barley, are common throughout temperate agriculture and also occur widely in tropical latitudes, usually at high altitude where it is cooler. Indeed, staples such as potato and some cereals originate from high altitude tropical regions (Chapter 4) and, although much land is cultivated using hoes and ploughs, the management of these crops in many tropical locations is little different to that in temperate regions. For example, cereal production on moist tropical plateaux deploys machinery, fertilizer and pesticide with an intensity similar to that in Europe or North America. Nevertheless, the principles advanced in this book should be relevant to a wide range of plants, husbandry and regions that fit the loose term 'tropical'. The examples described in the book are primarily of tropical species grown under low input husbandry, or in plantations, in low latitude regions of Africa, Asia and South America.

1.1 The boundaries, scale and scope of agronomy

Crop management, and its scientific study agronomy, are part of a system that comprises the physical elements of the climate, soil and land, the *biological* constituents of the vegetation and soil, the *economic* opportunities and constraints of markets, sales and profit, and the *social* circumstances and preferences of those who work the land (Fig. 1.1). In all this, management acts directly on a part of a plant, a whole plant or a small group of plants in a stand, or else an amount of soil that can be lifted or turned by a person, animal or machine. Each act of management influences the physiological processes of the plants, which in turn modify or regulate the flow of environmental resources – sunlight, water, nutrients – to

economic or useful products. Though large machines are used, especially in plantation agriculture and with intensely managed cereals, the animal-drawn plough is often the main means by which the soil is turned and shaped, while the hand-held implement, or the hand itself, sows, weeds, cultivates and harvests.

Despite being defined at a particular scale in this way, management and its effects influence and are influenced by processes and events at much smaller and larger scales. Any human influence on a plant organ affects the metabolic pathways that

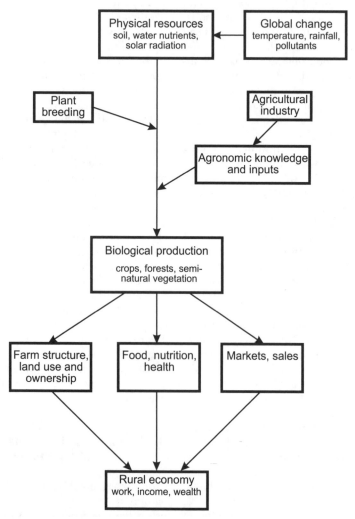

Fig. 1.1. Flow diagram of physical, biological, economic and social dimensions of agronomy.

give the organ its nutritional and biochemical character; and any disturbance to the soil ultimately alters its minute architecture and the organisms that inhabit it. Conversely, the climate, the topography of the land, and the economic and social factors such as markets and roads, determine the type of management that is possible and feasible. Many individual acts of management together influence the pattern of land usage, the fertility of the land and its local microclimate and the wealth and health of people and nations.

These interactions at larger and smaller scales than management are largely beyond the scope of this book. It is nevertheless essential that agronomists are aware of them. Historically, the physical, biological, economic and social disciplines each have had their own methods and units (or 'currencies') for which there seem no universally accepted exchange rates. We return to this theme in the final two chapters. The rest of this introductory chapter aims to summarize general aspects of tropical climates, soils and plant growth.

1.2 The broad dominance of climate

Climate is the long-term nature of the weather. Climatic variables that affect the growth of plants include solar radiation, temperature, precipitation, windspeed and atmospheric humidity. Other physical variables, such as the concentration of carbon dioxide and pollutants in the air, are also important to plants, but are not usually regarded as climatic, though, of course, they do influence changes in the climate.

The agronomist or farmer is not concerned with whether the field or region being studied is classified as being in the tropics. Attempts to classify what is tropical (e.g. Koppen, 1931, 1936; Paffen, 1967; Nieuwolt, 1977) can even mislead if taken too strictly. Koppen (1936), for example, demarcated the tropics as places having a mean monthly temperature of 18°C in the coolest month of the year. This definition is unhelpful as it is well above the limit at which most tropical plants start developing,

while large yields of tropical crops such as maize and tea have been won at temperatures well below it. Indeed, plants make their own distinctions, while farmers and practising agronomists do not have to be told whether they are in the tropics or not. In this book we are not concerned with precisely defining tropical climates, but with how different combinations of weather variables influence the growth and yield of crops.

Any simple definition of tropical climates is complicated by the effects that latitude, altitude, topography and atmospheric circulation have on a general pattern. As a consequence of this variation, local climates within the boundaries of the tropics can vary by as much as anywhere in the world. Strictly, the lines of the Tropics of Cancer and Capricorn, respectively 23.5° N and 23.5° S of the equator, indicate the outer limits where the sun can ever be directly overhead. Between these limits and some way beyond them, the solar radiation is very intense. Here, leafy crop plants may have to transpire water quickly to stay near to the surrounding air temperature or else they will scorch. However, at these latitudes rainfall may be sporadic and amounts may be small. The water held in many tropical soils must therefore enable stands to remain alive for several weeks and sometimes several months without significant rainfall. At lower latitudes, i.e. closer to the equator, rainfall amounts often substantially exceed those at the margins of the tropics and rarely constrain the growth of vegetation. Between these extremes, topography and particularly altitude can influence the pattern and availability of rainfall over extremely short distances.

1.2.1 Climate and vegetation

The broad type of vegetation in a region is strongly determined by its climate. Even where human activity has greatly altered the composition and extent of plant cover, the local climate dominates all other factors in determining where, when and for how long particular crops can be grown. For a given species, the climate sets the limit to total

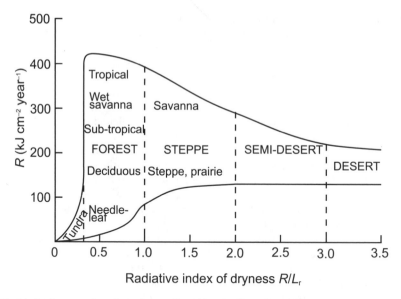

Fig. 1.2. Budyko's diagram of geobotanic zonality. (After Budyko, 1974.) The parameter R/L_r defines the relative components of the heat and water balance (i.e. a dryness index) and R is the heat balance. The principal geobotanic zones are then demarcated along this axis as straight lines. Essentially, we can consider the vertical axis as representing heat and the horizontal axis as a measure of wetness.

productivity and yield. Even in highly developed agriculture, climatic factors still alter the annual yields of the main species substantially between one year and the next. In addition to direct effects on the primary production of vegetation, climatic factors also affect the performance of plant systems by changing the populations of pests, diseases and competitors and making soil more or less difficult to till or work on.

Because other climatic factors change less dramatically over short distances, rainfall is usually the most dominant factor controlling the type of vegetation and agriculture and is the basis of many classifications of climatic zones (Thornthwaite, 1948; Troll,

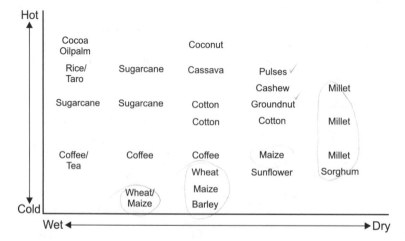

Fig. 1.3. A simplified matrix of the major crop types in relation to climate.

1964). Budyko's classification (1974) also, in effect, includes evaporative demand through an index of dryness, which depends on the difference between incoming and outgoing energy and on the rainfall. From measurements at 1600 sites throughout the world, he was able to assign a range of the index to a broad type of vegetation – forest, steppe, semi-desert and desert (Fig. 1.2). We can take the basic characteristics of Budyko's classification of *natural* vegetation and create a simplified matrix of major *agricultural* vegetation, i.e. crop types that typically fit within the constraints set by radiation and evaporative demand (Fig. 1.3). The distribution of most rainfed crops is within the comparable zones of forest and steppe. Outside these zones, farmers without access to irrigation must select crops that are highly resilient to the stresses imposed by shortages of soil moisture, often accompanied by high temperatures during much, if not all, of the growing season. The physiology of the species and skill of management determine the actual ranges of crop plants within these zones. (Examples are given in Chapters 3, 4 and 5.)

1.2.2 Climate and wealth

From the above it can be seen that the local climate can determine the affluence of human communities and ultimately the wealth of nations. For example, Fig. 1.4 shows the distribution of the least developed countries of the world as defined by the United Nations (1985). At the time of that report, the majority of the world's poorest countries were in a contiguous sequence across the Sahelian region of Africa, extending south to Tanzania, Malawi and Botswana. Each country is within a region defined as arid or semi-arid which roughly corresponds to the desert and semi-desert zones defined by Budyko (see Fig. 1.2). Here, although there is little diversity in the natural ecosystem, agricultural products still account for the bulk of national income. In recent years, this group of countries has mostly shown stagnant or declining agricultural productivity with only modest increases in the value of their export crops. Most of the other poorest countries in the world are in regions of high altitude where poor, thin soils and topography, in addition to constraints of climate, make agriculture difficult or impossible.

At the other extreme, much of Bangladesh can receive more than 3750 mm of annual rainfall, often in torrential downpours accompanied by strong winds. Much of the country is barely above sea level and storm waves and sea surges raise the water level along the coast so that widespread flooding of low-lying areas occurs, adding to

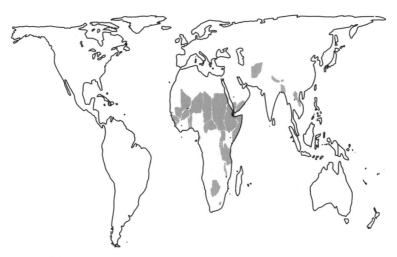

Fig. 1.4. Geographical distribution of the least developed countries defined by the United Nations (1985).

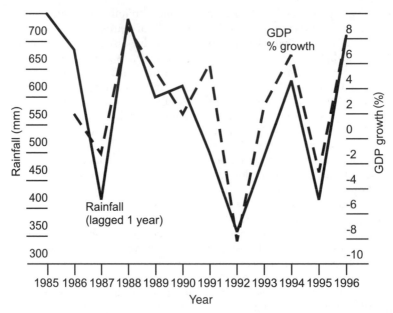

Fig. 1.5. Growth in the gross domestic product (GDP, %) of Zimbabwe against previous year's rainfall (mm).

the devastation caused by the strong winds. A major element of agriculture in these regions is the uncertainty associated with the timing, amount, intensity and duration of rainfall. Thus, irrespective of the many social and political constraints on them, the economies of the least developed nations ultimately depend on a vulnerable, uncertain and increasingly impoverished natural resource base. It is the declining productivity of this base that is the underlying cause of their poverty.

The dominance of a single climatic factor, in this case rainfall, on the economy of an entire country is demonstrated in Fig. 1.5. Here, despite the inevitable economic complexities and social pressures imposed on a newly independent nation, the annual gross domestic product of Zimbabwe between 1986 and 1996 was closely correlated with the previous year's rainfall.

1.2.3 Climatic variables and crop production

The climate at any one place is the product of year-to-year variations in daily and seasonal weather patterns. These patterns are a consequence of changes in solar radiation and the atmospheric systems that control cloud cover and rainfall. The exact effect of each climatic variable is often difficult to define because weather variables are highly correlated and steady state conditions are rare in the field. However, seasonal trends in the principal climatic factors can be quantified in terms of the latitude, altitude and geography of a particular site. For example, Table 1.1 describes the mean winter and summer values of irradiance, daylength and temperature at sea level for three latitudes in the northern hemisphere.

In terms of their influence on crop processes, the major climatic factors interact between two atmospheric systems. Directly or indirectly, the *solar spectrum* controls the solar radiation, daylength and temperature at a particular site. The *terrestrial hydrological cycle* determines precipi-

Table 1.1. The total solar radiation, daylength and mean temperature at sea level for three latitudes in the northern hemisphere. (Adapted from Woodward and Sheehy, 1983.)

		Latitude (°N)		
		0	20	50
Winter				
Mean irradiance	(W m^{-2})	411	296	87
Daylength	(h)	12.1	10.9	8.1
Mean temperature	(°C)	29	19	−1
Summer				
Mean irradiance	(W m^{-2})	417	451	470
Daylength	(h)	12.1	13.4	16.4
Mean temperature	(°C)	27	32	18

tation and evaporation. The interaction of temperature and evaporation controls the atmospheric saturation deficit above a crop. Saturation deficit is a measure of the moisture content of the air and it is this factor which determines the atmospheric demand for evaporation at any particular location.

Figures 1.6–1.8 illustrate how the solar spectrum and hydrological cycles influence the annual totals and seasonal distributions of the major climatic resources at each latitude. These totals and distributions set limits on the type of vegetation that can be grown at any location and at which time of year it can successfully grow. Figure 1.6 shows the annual trend in daily solar radiation received at four locations between 1 and 64° N of the equator. Interestingly, the peak value of radiation received during the year at Reykjavik is actually greater than that

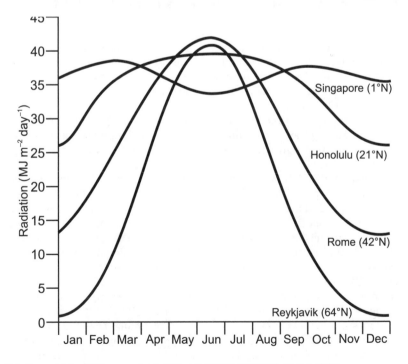

Fig. 1.6. Extraterrestrial radiation received at four locations (latitude) on the globe.

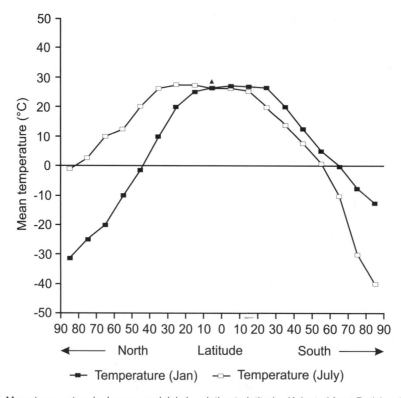

Fig. 1.7. Mean temperature in January and July in relation to latitude. (Adapted from Budyko, 1974.)

at the equator. This is largely because at high latitudes longer daylengths in the summer compensate for the lower intensities of radiation. However, because the daylength at the equator varies little between summer and winter, the radiation received remains similar throughout the year and the annual receipt of radiation therefore greatly exceeds that at higher latitudes.

The seasonal distribution of solar radiation determines the temperature and rainfall patterns at any location. Figure 1.7 shows the variation in mean temperature in January and July with latitude. At any time of year, there is convergence towards the equator. However, as with solar radiation, the annual range of temperature increases with distance from the equator.

All atmospheric moisture originates from the earth's surface, where water in liquid or solid phase is transformed into water vapour via the process of *evaporation*. There are two components of evaporation. The first

is that which occurs *directly* from water surfaces or from land areas which have soil or vegetation cover that has recently been wetted by precipitation. The second source is *transpiration* which is physically the same process as direct evaporation but occurs via plants which draw water through their root systems. Direct evaporation and transpiration are often combined into the collective term 'evapotranspiration', E_t, which is a measure of the total flow of water vapour to the atmosphere. The forcing effects of solar radiation and temperature, demonstrated in Figs 1.6 and 1.7, mean that the latitudinal variation in annual evaporation increases at lower latitudes (Fig. 1.8). However, maximum evaporation is not centred at the equator but reaches two peaks located at about 25° N and 15° S of the equator.

The annual evaporation at any location is a measure of the demand of the atmosphere for water vapour. In most cases, the principle means of satisfying this demand is

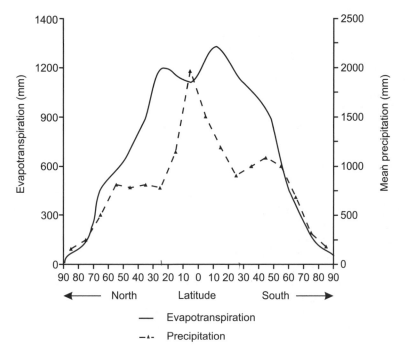

Fig. 1.8. Latitudinal variation in mean annual evapotranspiration and mean precipitation. (Adapted from Wust, 1922; Sellers, 1965.)

through precipitation, usually in the form of rainfall. Figure 1.8 also shows the latitudinal variation in precipitation across the planet. Although the shape of the precipitation distribution shows a crude similarity to that of evaporation, the two curves are not congruent. In fact, the previously mentioned dip in evaporation at the equator coincides with the increase in cloud cover which is itself a consequence of the maximum values of precipitation. These complex interactions mean that there can be large differences in the balance between the supply (precipitation) and demand (evaporation) for moisture at different locations on the planet. For example, whilst locations at about 55° N and 25° N each receive approximately 500 mm of annual precipitation, there is an approximate doubling of evaporative demand across the same latitudinal range. As a consequence, the supply and demand for water is broadly similar at 55° N (e.g. temperate north-west Europe) whilst at 25° N (e.g. semi-arid Africa), the annual demand is roughly double the supply.

Of course, the general links identified above between major climatic factors, such as radiation, rainfall and evaporation, and latitude are a gross oversimplification of the real world. Other factors such as topography and, in particular, the atmospheric circulation play a crucial role in determining the exact climate at any specific location.

Atmospheric circulation

The local climate and weather that crop management has to contend with in any year or field are determined by global factors. The poleward flows of air from the tropics, caused by the uneven heating of the earth's surface, are the main driving force for the longitudinal air currents of the tropical circulation (Lockwood, 1974). Combined with the rotational force of the earth, they cause air to move and produce large convection currents which determine the global pattern of winds. At the equator, the circumference of the earth is about 40,000 km, and any point on the equator is moving eastwards at just over 1600 km h^{-1}. In contrast, at the North Pole, there is no such movement. Air, which is not itself moving, has a true eastward velocity and momentum that is similar to that at the surface. When this air mass moves to a different latitude, it will retain some of its original eastward momentum, though much is lost with friction at the earth's surface. As the air moves away from the equator, its excess eastward momentum relative to the earth's surface makes it appear as a west wind to an observer on the ground.

The latent and sensible heat of the moving air masses that originate in the tropics are transported by the winds and gradually released to colder air. The poleward moving air currents (anti-trade winds) generally decay at 20–30° latitude. Here, the masses of air they contain fall back to the earth's surface, resulting in areas of relatively high atmospheric pressure (sub-tropical highs). This circulation is completed by a massive air stream drawn back towards the equator (trade winds). The two trade wind systems, each side of the equator, converge at an area called the Inter Tropical Convergence Zone (ITCZ) which is characterized by relatively low pressure and rising air movements. This generalized global circulation is described in Fig. 1.9.

The simplified global pattern shown in Fig. 1.9 is complicated by the seasonal differences in radiation received in tropical and sub-tropical regions some distance from the equator. Latitudes between 15 and 30 degrees receive very large amounts of radiation during the summer months. The main consequence of the resultant heating of the land surfaces at the beginning of summer is that the subtropical high pressures are transformed into centres of low pressure over the tropical continents. As a result, a new area of intertropical convergence is established some distance from the equator. The tropical continents and adjacent oceans therefore experience a temporary reversal in wind direction, and the winds in these systems are known as 'monsoons'. The layer of air in which most of this circulation occurs – the 'troposphere' – is shallow, with a maximum depth near the equator of about 20 km. In contrast, its horizontal dimensions are measured in thousands of kilometres. Because of this shallow and wide character, horizontal air movements (i.e. 'winds') pre-

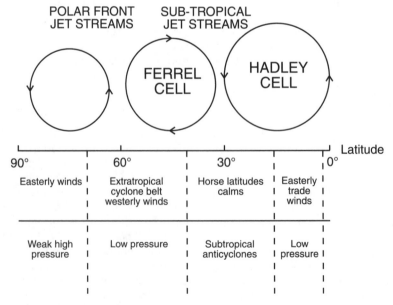

Fig. 1.9. Mean meridional circulation of the northern hemisphere. (Adapted from various sources including Budyko, 1974; Nieuwolt, 1977.)

vail. However, vertical movements are also extremely important because they transport heat and moisture to and from the earth's surface and change the stability conditions of the air masses involved.

The monsoonal systems dominate the climate in much of tropical Asia, Africa and South America and deliver most of the rainfall at latitudes between 15 and 30°. The consequences of monsoons for vegetation and people are enormous. Climates controlled by monsoons include more than half the world's people, most of whom derive their income almost entirely from agriculture. Yet within all three continents, there are large variations in rainfall and its associated climatic effects. In Asia, parts of Malaysia, Indonesia and New Guinea receive rainfall throughout most of the year; the Indian sub-continent, Burma, Thailand and across to the Philippines have a dry-and-wet monsoon climate, with wet summers and cooler dry winters; and there are parts of the area to which monsoons fail to bring rainfall, as in the north-west of the Indian sub-continent. In tropical Africa, there is one monsoonal system in the west and another in the east, but large areas in the centre and south have no monsoon but experience winds of the general circulation. Much of the interior is highland plateaux,

generally at elevations above 1000 m, which gives the interior a highland continental climate. Seasonal rainfall shows a distinct bimodal distribution, with two maxima near the equator and a single maximum at latitudes greater than about 10 degrees. In tropical America, the variations in climate are also very great, but more systematic, with generally wet climates to the east of the central mountain range, and dry ones to the west. There is also a very great change from the north of South America, where annual rainfall is over 5000 mm, to the deserts of the south.

The annual distribution of rain by these circulation patterns is the main factor that sets the start and duration of agricultural activities. At any location, it is not only the *amount* of rain but its *variability* within and between years that affects agricultural development. When the rain does fall, it is often in intense storms that erode the soil or compact its surface or in such small amounts as to be inaccessible to plant roots. The variability of mean annual rainfall also increases as the climate becomes more arid, such that rain is least reliable where it is most needed, increasing the risks associated with crop production.

Sometimes tropical regions suffer a sequence of bad years in which rainfall is

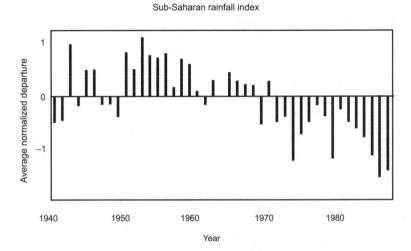

Fig. 1.10. Rainfall index for 20 sub-Saharan stations in West Africa west of 10° E between 11° N and 19° N. (Developed by Lamb, 1985.)

below the long-term average. An example of this is the drought in the Sahelian region of West Africa which lasted for 17 years from 1968. Figure 1.10 shows data obtained from 20 sites in West Africa, west of 20° E and between 11° N and 19° N (Lamb, 1985), and demonstrates how rainfall can decline drastically over a sequence of years. The changes had devastating consequences for the inhabitants of the region. There are also recorded instances of changes in seasonal weather that have lasted much longer than this. In eastern Kenya, for example, a new 3-month long dry season appeared in 1913 which has lasted ever since (Musembi and Griffiths, 1986), and to which agriculture must have been forced to adapt.

1.3 The influence of soil and topography

At scales below tens or hundreds of kilometres, the shape of the land, the state of the soils and agricultural history have an influence at least as great as that of climate. Steep slopes, marshy ground and other such features severely limit the types of management possible, unless overcome, for example by terracing or drainage. Between extreme features, though, agriculture grades from one form to another. Gradients of temperature, dryness, soil depth and fertility might run together or in opposition in different parts of the world to create zones where one form of agriculture or another predominates. A typical transition in Africa, for example, is from a dry plane or valley bottom supporting extensive grazing, through a range of mixed animal and crop systems on wetter slopes, to vegetative crops such as timber and tea in cool wet uplands.

More local features can blur the boundaries between zones or create extremes of zonation over short distances. This is so even in the humid tropics. In south-east Asia for example, the common transition in the natural vegetation from coastal high forest on deep wet clays and peats to low montane scrub on thin lateritic soils is reflected in the yields of crops that replace the natural vegetation. In extreme instances, a change from

mangroves and swamp forest to seasonal grassland with xerophytic and insectivorous plants can occur over two or three kilometres, largely as a result of change in soil depth from several metres to a few centimetres. Such differences in vegetation are maintained under one climate.

Whether the transitions occur rapidly or slowly over distance, the range of crop species that can be grown is still determined by physiological interactions of plants with soil and air. Vegetative crops predominate at high, cool regions of the tropics because low temperature stops reproduction in most tropical plants before it stops their foliage expanding. Similarly, determinate cereals are rare on the dry plains because more often than not they scorch before grain is set and ripens. Selection, breeding and agronomy can all extend the range of a species, for several years at least. Over a longer time, an agricultural species can be continually grown in an environment only if the nutrients removed from the soil by the crops are replenished. Further information on physiological differences between species is given in Chapter 4 and on sustainable management in Chapter 8.

1.3.1 Heterogeneity of the soil

Zones of specified climate, altitude and terrain are themselves not uniform and can only guide possible agronomic practice. Patterns on a smaller scale of hundreds of metres occur in both the natural vegetation and as a result of earlier agricultural activity. In the Amazonian forests of Bolivia, for example, log jams in rivers start a sequence of events that alter the natural flora. Flooding and deposition of river-borne soil caused by the log jams leads to death of forest trees and temporary clearings that are then colonized by trees such as broad-leaved mahogany (*Swietenia macrophylla*), which normally are unable to grow in denser forest. Distributed throughout a catchment, many such events create a heterogeneous vegetation and a shifting system of tributaries. Removal of the natural vegetation in parts of the Amazon has also revealed evidence of past agricultural activity now obscured from aerial view by forest. The species composition of the forest will be influenced by this previous agriculture and any future agriculture in this and similar forest will undoubtedly be influenced by previous ecological change in the distribution of species. In such ways, agriculture and the natural environment are commonly interdependent and interchangeable. It is logical to consider agricultural management within the context of an *agroecosystem* in which similar biological and physical principles can be applied to both natural and managed species. A wider discussion of these principles is presented in Altieri (1995) and Collins and Qualset (1999) and we return to this subject in Chapter 8.

Localized changes in soil physical properties that result from weathering of the underlying rock and from soil erosion can be enhanced by biological activity. Systematic change over metres and tens of metres in the physical and chemical nature of the soil commonly occurs from the movement of soil particles and nutrients down slopes. Even on flattish land, complex mosaics have developed from the underlying chemical nature of the soil. In the Kenyan savannah, for example, the patterns of grassland communities were mainly related to sub-surface sodium concentration and mound-building termites (Belsky, 1988).

Arable agriculture inherits this heterogeneity, is forced to adapt to it and tries to modify it. Some of the effects of spatial variability can be erased by using machines to plough and smooth the land, by applying fertilizer and irrigation and by growing crops that are genetically uniform or homogeneous. McBratney (1992) presents results which show that the great spatial variability in pH at 0.2 m below the surface is much reduced in an adjacent paddock. However, even when subjected to intense agriculture for several decades, some spatial patterns are very resilient and difficult to eliminate. Often spatial variability over scales of 10–100 m also persists in the populations of nematodes and other invertebrates in many fields.

To the farmer or researcher, natural and imposed variations in the structure, consistency and biology of soils and resultant variations in the growth of plant roots are often disguised. This is because, unlike the aboveground environment, they are hard to see. Whereas the aerial environment of a crop stand can be relatively well described by measuring the fluxes of radiation, temperature, vapour concentrations and windspeed, etc., the vertical and horizontal fluxes of physical and biological elements within the soil are latent and are, therefore, more difficult to measure. In many research environments, efforts have been made to ameliorate natural variations within the soil to try to achieve uniformity for experimental purposes. In some instances, attempts have been made to create 'precision' fields by using machinery to redistribute soil and provide a uniform and flat surface. It is our experience that, even when such drastic and expensive procedures are attempted and do indeed produce a visually uniform surface, they are doomed to failure because the physical redistribution of soil rarely achieves or maintains a uniform composition of structural or biological elements with depth. With soils, at least, what you see is rarely what you get.

Whereas highly mechanized management attempts to achieve homogeneity of the soil, farming on infertile or otherwise difficult soils with few inputs often gains a yield by enhancing spatial variations within the soil. Slash-and-burn agriculture, water-harvesting, raised beds all attempt to concentrate water and nutrients in particular places in relation to crop plants. This imposed heterogeneity can remain for considerable periods, as evidenced by the persistence of field structures from ancient times. However, not all of this heterogeneity is persistent. Agronomists in Malaysia had difficulty establishing oil palm on peats, because once drained, the organic matter within them oxidized, the soil level dropped and the palms fell over. Planting the palms in a hole to an extent solved the problem because after several years the soil surface moved down to be level with the base of the plant.

This type of imposed heterogeneity

commonly occurs over centimetres to several metres and is usually visible in the nature of the soil surface or among the plants themselves. It is on smaller scales, however, that most tropical management interacts directly with the heterogeneity of the soil. The spade and plough more generally affect crucial processes at scales of millimetres and decimetres: those at which channels form through which water and gases move, roots explore and exploit, and bacteria, fungi, protozoa, nematodes and other invertebrates graze and predate. Some typical scales at which different processes work in the soil are shown in Fig. 1.11.

The fine structure of soil

Particularly important for plants are the actions of microorganisms. Some of these are closely associated with the plants themselves. The nitrogen-fixing symbionts that live in the roots of legumes and certain other plants contribute nitrogen, while the fungi that combine with roots to form *mycorrhiza* can enhance the uptake of other elements. Besides these, the fertility of soil depends on many free-living bacteria and fungi, which break down organic matter and take up, store, transform and release nutrients, and whose secretions and dead bodies give cohesion and structure to the soil. Structure is important to enable a soil to hold a diverse assemblage of organisms and to resist the power of rain and wind to erode soil particles and wash out nutrients.

Many soils are able to support a community of microorganisms having a much greater mass and containing more nutrients than the crops themselves. The microorganisms compete with plants for the nutrients, which they lock up for a time, but then release back to the soil or to the atmosphere. One of agronomy's main aims is to understand how to optimize rates of amassing microorganisms in stable communities and of releasing nutrients when plants can best absorb them. One of husbandry's greatest skills is to do this blindly, without any immediate information about microorganisms or nutrients.

Implications for agronomy

Agronomy and husbandry work blindly at the fine scales of the soil pores and chan-

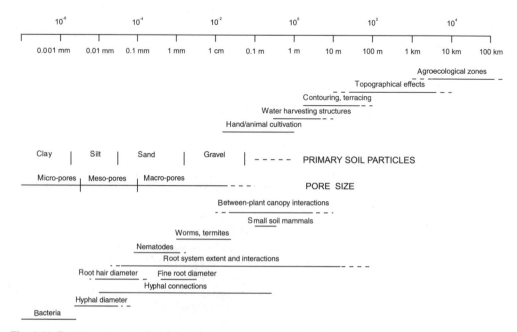

Fig. 1.11. Typical scales at which different processes work in the soil.

nels. To the experienced, the success of a practice can be judged by the feel and workability of a soil and the yield of crops over several years. To a large extent, science is blind here also. There are enormous difficulties in understanding links between the microscopic nature of soil, tillage and yield. Moreover, the synthesis of knowledge above and below ground has been hampered by traditional divisions in research and teaching that still persist, and in some cases are being reinvented. At present, science knows perilously little about the effect of soil structure and function on the way plant stands establish, absorb resources, grow and yield. Accordingly, quantitative descriptions or mathematical models of soil and plant tend to simplify, and often misrepresent, the processes concerned. We know, for example, that soil is not a homogeneous mass. It often has cracks and channels through it caused by worms, termites and moles or contraction during drought. One of the authors remembers pouring several buckets of water down a 2 cm wide crack in a droughted lateritic soil: the crack showed no indication of being filled with water. Following a dry season, rain washes down such cracks for a time, possibly leaving much of the bulk soil dry. The reality, however, is more as represented in Fig. 1.12, where many channels of different size penetrate a matrix of fine and tortuous pores through which water, gases, nutrients, microorganisms, roots and invertebrates are unevenly distributed.

Because the fine structure of soils is so complex, many models and equations still treat soil as if it were homogeneous. As a first step, this provides the student with the main soil factors that can influence crops. However, an important result of soil heterogeneity is that the value of a soil factor usually changes with the scale of measurement. The rate at which water infiltrates soil, for example, will often increase if a measuring device of larger diameter is used as larger and larger pores and holes become encompassed by the device. Similarly, at an invisible scale, bacterial processes might have a good supply of oxygen if they occur on the outside of soil aggregates but be starved of oxygen if they occur deep inside fine pores, only a few millimetres away. Interpreting soil function through an average or bulk value can therefore be misleading. For example, a soil at one bulk measure of oxygen concentration will likely support both anaerobic as well as aerobic processes.

There is growing interest in whether heterogeneity and its effects at all these regional and microscopic scales can be interpreted through a common set of principles. We return to this theme in the final chapter. In Chapters 3 and 7, we present some simplified guides to the processes linking soils to plant growth. Farmers and agronomists know that the reality is more complex, but it is hoped the simple concepts will help to initiate more intensive and directed agronomic experiments, less demanding of time and land, as well as attempts to understand heterogeneity from a common starting point.

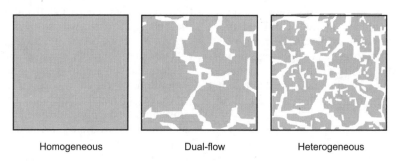

| Homogeneous | Dual-flow | Heterogeneous |

Fig. 1.12. Simple representations of soil structure (diagram provided by Dr B. Marshall).

1.4 Climates, crops, soils, people

One of the biggest challenges facing the tropical agronomist is to link the various physical and biological processes that set the potential and determine the actual performance of a diverse range of crops and cropping systems with the human dimension involved in crop management, preference and choice. A logical starting point is to define the environmental conditions within which plants and humans have to operate.

1.4.1 Agroclimatic classification

One response to the unruly diversity of agriculture has been to classify it. The system of agroecological zones in Kenya (Jaetzold and Kutsch, 1980; Jaetzold and Schmidt, 1982) is based on soil, topography, plants and animals as well as climate. Their classification is a matrix of rows and columns similar to the simplified scheme depicted in Fig. 1.3.

Land classifications were originally done by assimilating knowledge from surveys and estimates provided by regional agricultural departments. Ground surveys can now be augmented by analysing photographs from aircraft and satellites. The balance of different wavelengths in the spectrum of visible and infrared radiation is sometimes used to assess the developmental stage and health of vegetation. Caution is needed, however, over received classifications of agricultural land and soil. Some soil scientists and agronomists recognize 'cotton soils', 'cocoa soils', 'rubber soils' and other soils that they recommend specific crops to be grown in. Some national organizations have even produced maps showing where species should be planted. Land use types and categories might be useful for legislating against malpractice or encouraging husbandry that stabilizes slopes, for instance, but often they do not correspond with what is growing or will be grown in the ground.

1.4.2 The reality of infrastructure, inputs and profit

While climate, soil and topography together determine the potential range of a plant species, and the range of species that can be grown at any site, other factors contribute to, and sometimes dominate, what is grown at a site. Chief amongst these are the availability of plants, the desire to grow rich off the land and the need to appease hunger.

Early trade, human migrations and colonialism moved plants from one continent to another. South-east Asia, for example, has an ideal climate for perennial tree crops, but massive and profitable plantation agriculture developed there only after rubber was brought from South America and oil palm from West Africa. Potatoes from the Andes became a staple diet in parts of Europe (hence the 'Irish' potato) and are now found in cool tropical regions throughout the world. Maize from central America feeds most families in Africa. Given that few barriers now exist to the movement of crop species, the range of crops and products in cash agriculture, and the resulting pattern of land usage, often depend as much on policy, economics and preference as on the physical environment (see Chapters 4 and 8).

Tea, for example, is cultivated in broadly similar cool-to-warm moist climates at a range of sites throughout Africa and Asia. It is also grown at hot, moist, previously forested lowland sites in Assam and Sri Lanka, but only in small quantities at similar sites in Malaysia and hardly at all in many other similar climates. Tea could be grown at most of the hot, moist sites in the world, but is often simply not as profitable there as rubber, cocoa or oil palm. During the 1980s, oil palm was so profitable as a crop in south-east Asia that it was grown in preference to all other crops on all types of soil. Similarly, a single species – sugar cane – dominates the landscape of Mauritius regardless of terrain and local climate. In subsistence cropping, also, factors other than those of the climate and land contribute to profitability and success. Particularly important are the availability of inputs such as fertilizer, whether farmers can afford to buy them, their cost in relation to the return they bring, the costs of transporting produce to the market, the selling prices and the associated subsidies, if any. Most importantly, despite any analysis or prediction based on physical and economic factors, many crops are grown in preference to others simply because farmers and their families know how to grow them and like to eat their products. The lesson is always to look farther than the local physical environment to cover all possible reasons why a crop is where it is and how many human and physical resources are expended on it.

1.5 Agronomic experiment and analysis

Agronomy, as any other biological field of study, is faced with understanding complex processes. It also has to synthesize knowledge from many sources and disciplines to solve immediate and practical problems. Agronomy can rarely achieve understanding at scales of the farmer's stand and field because of the many couplings and feedbacks that exist between weather, soil and plants and management. If one factor is changed, several others change also. Agronomic research therefore developed a strong experimental base.

The essential feature of biological and agronomic experiments is to break some of the couplings between variables. For instance, plants might be grown in controlled environments where temperature is varied but water held constant, or stands might be grown in small plots over a range of planting density, each receiving the same amount of fertilizer. Such experiments tighten the 'boundaries' of the system: if they make temperature vary, for example, they will attempt to keep within defined bounds for the degree of variation in saturation deficit. The problem is then examined within the new boundaries. The ultimate aim is to move back to the original problem in the field, and to do this, the true effects of the broken couplings have to be inferred from what was learned about the processes (Squire and Gibson, 1997).

In reality, it is very difficult to make this move back to the field. Problems tend to be tackled at different scales by people from different scientific disciplines and codes such as soil science, physiology and genetics. When the disciplines have remained isolated, problems at the field scale have rarely been solved. Even when disciplines combine, there are still fundamental scientific challenges of scaling, in which some mental or mathematical procedure is used to predict from measurements in experiments what is likely to happen at the scales of crops in soils subject to variable weather and heterogeneous soil (Marshall *et al.*, 1997). A further complication is that the tools of mathematical interpretation are not always familiar to the biological practitioner or researcher and provide another barrier to understanding.

In some areas of science, so little is known about couplings at the scales of the population, field and catchment that the inferences can only be guesses. Predicting effects of global environmental change on vegetation, for example, is very uncertain because experiments are not feasible at scales anywhere near to those of the question. The fate of introduced genes and cultivars also presents particular problems of scaling. The initial change occurs at the scale of the individual genome or small section of DNA. The ramifications might make millionaires, destroy a country's soil or contribute to a stable equitable agriculture.

Agronomy itself is fortunate in that experimentation at near-realistic scales is possible. Indeed, agronomy has generated a range of experimental and theoretical methodologies, among which are empirical field trials, physiology and environmental physics.

1.5.1 Empirical field trials

A multi-factorial field trial is put in place to get a response curve at the field scale to a variable such as fertilizer or planting density. The purpose of such trials is to draw out effects of treatment from those of local topography, nutrient status and other soil factors. The treatments are repeated in a number of blocks sited in different parts of the experimental field. A branch of statistics has been developed to analyse the results of these experiments (see, for example, Lane *et al.*, 1987) and assess their validity.

An end-point, such as yield or the quality of yield, has often been the only response quantified. Nevertheless, trials coordinated throughout a country or region can provide a description of the broad effects on yield of soil type, climatic variation and nutrition. The coordination of disparate fertilizer trials in Malaysia by the national oil palm plantation body is a good example of this approach (Foster and Goh, 1977).

Scaling-down field trials has been tried to obtain, in a much smaller space, a response curve of yield to nutrients or planting density. Examples are *fan designs*, developed for vegetables (Nelder, 1962; Bleasdale, 1967) but applied to crop plants as large as oil palm (Corley *et al.*, 1971a), and *line-source* irrigation experiments (Hanks *et al.*, 1976). The aim of all miniature trials is to establish a gradient of the variable being studied. Plants at any single point on the gradient might extract additional resources at higher or lower points, so scaling to the level of the commercial field is not straightforward. Nevertheless, the small, response-curve trial can give a good first indication and breadth of effect generally and, moreover, is at a scale quite appropriate for mixed gardens and uneven terrain where steep gradients in fertility and soil depth are commonplace.

Empirical field trials have remained the stock-in-trade of tropical agronomy for decades. The working doses of fertilizer and irrigation in many crops have been defined from the results of such trials. They have limitations, however. They are repetitive and need much land, money and time. They rarely identify links between the physical environment and yield, so are site-specific: the results at one site can seldom be applied elsewhere. Moreover, the scale of the plot of plants whose yield is measured and that of the guard area around them have to be representative of the real conditions under which the crop will be grown. Plantation crops such as oil palm, rubber and tea covering vast stretches of land in a more or less unbroken blanket, have very different edges from the single rows and clumps of plants in subsistence gardens. The purpose of the guard area is to minimize the effect of the edge of the experiment or the edge of another treatment. Plants at an open edge get more light and nutrients than those nearer the centre of a plot, but they also lose more water. Therefore, they might grow faster or slower than the centre ones according to the circumstances. As a general guide: the larger the extent of a crop, the more important it is to remove the edge effect in any trial. So small plots of wheat or maize in breeding trials, for example, that are grown without guard areas will yield much more than they would in an intensely managed field, simply because individual plants in the plot receive more light. Scaling these results upwards would lead to an overestimate of the yield from a large field. Scaling results down from large trials would likely underestimate the yield of single rows or single plants.

Replicated field trials will remain an important feature of agricultural research for many years, especially where they are practicable and affordable. The branches of agronomy that seek mechanisms that link weather, soil and crop have the important functions of extending the results from trials, finding alternative short-cuts and eventually removing the reliance on these trials by many tropical research stations. They also contribute to an understanding of cause-and-effect which, as well as explaining what we already can see, allow us to predict the future performance of crops and cropping systems in uncharted territory. In this context, modern techniques, such as the use of computer simulation models of crop growth and yield have an important role to play in furthering our understanding of what we know and, perhaps more importantly, what we don't know about particular crops and cropping systems. We return to this theme in Chapter 8.

1.5.2 Physiology in controlled environments

To understand mechanisms linking plants with environment, a branch of agronomy retreated to controlled environments where the heterogeneity and couplings of the outdoors could be eliminated. In the laboratory, agronomy joined with basic physiological science to investigate the independent effects of light, water, nutrients, temperature, photoperiod and other variables. Accordingly, several crop plants such as maize, wheat and barley became used widely in laboratories as objects of basic physiological research.

Experimentally restricting the couplings among processes and environment often brings about untypical responses in plants (Raper and Downs, 1976). Low radiation in chambers has so often limited the carbohydrate for plant growth that the response to other variables has simply not occurred. Altering temperature while allowing humidity to vary unchecked is another common limitation, as is maintaining a given relative humidity over a range of temperatures rather than a saturation deficit (see Chapter 3). Some experimenters therefore moved to systems that did not rigidly control all variables. Open-topped chambers and controlled-environment glasshouses, for example, allow sunlight to vary more or less naturally while still regulating factors such as temperature and carbon dioxide concentration. Others went further to maintain that, since plants rarely experience a constant environment, daily variation and 'noise' in factors such as sunlight, temperature and saturation deficit should be included within the experimental treatment. Moreover, they insisted that plants' roots and canopies should be allowed to form without being severely restricted by their containers (Monteith *et al.*, 1983).

Physiological studies in controlled environments have made important contributions to understanding crop growth in the field, but only for specific processes. As a general guide, the responses to environmental factors that exert local effects, and do not have to operate through a whole-plant growth rate, have been reasonably well quantified by physiological experiment. Many effects of temperature and photoperiod on rates of leaf initiation and floral development are of this type (Chapters 2 and 5 give examples). In contrast, those effects of sunlight, water or nutrients, for example, that operate through a whole plant or whole stand have often been misrepresented in controlled environments. A common error – as already indicated – has been to observe no response to varying one factor, such as nutrient concentration in the medium, while another factor such as light is limiting the rate of growth. Agronomy should not assume that the variable factor exerts little or no effect in the field.

1.5.3 The reality and uncertainty of the field

A parallel approach of plant science was to measure and attempt to interpret yield through processes of growth and development. Early fieldwork yielded growth curves in different environments, or population structures of species in competition (e.g. Harper, 1977). In agriculture, the direction of this approach became concerned less with the variability within populations than with the dimensions of structures that make up a stand. A central analysis developed in which the performance of a stand was expressed in relation to some measure of its size or surface area (Rees, 1963). The *relative growth rate*, for example, is the dry matter increment over a period divided by the dry matter at the start of the period; while *net assimilation rate* is the dry matter increment divided by the leaf area index.

This form of expression provided a common means of defining growth, but often led to misunderstandings. Expressing the rate at which new dry matter is produced in relation to plant mass, for example, means that even if the absolute rate were stable – or even maximum – the relative rate would tend to decline as plants become larger. A declining rate during the main phase of crop growth might suggest that something is wrong. By contrast, a declining relative growth rate and stable absolute growth rate are common in healthy crops once canopies intercept most of the available light (Squire and Corley, 1987).

Growth analysis in the field suffered some of the limitations of empirical trialling. In particular, the results from any site, season and crop were not transferable to other environments. A stand of a given leaf area index (i.e. the (dimensionless) leaf area per unit field area) might grow at a slower rate or for a shorter time at one place compared with another, but the methodology was able to give few insights as to why factors such as leaf area index were not governing absolute growth rate between environments. These insights came with the merging of physical and plant sciences over the last few decades.

1.5.4 The contribution of environmental physics

Dimensions such as leaf area index became more directly linked to the energy in the environment following the concept of potential evaporation, which was an estimate based on physical theory of the maximum rate that water could be lost from vegetation (Penman, 1948). Hutchinson *et al.* (1958), for example, used the concept of potential evaporation to estimate water use by cotton in Sudan. They derived, experimentally, a relation between leaf area index and evaporation as a percentage of potential evaporation. They also brought to the field the concept that the onset of water strain during the season could be detected by methods of relative turgidity developed by physiologists in the laboratory. Their analysis accounted for much of almost a tenfold variation in yield among planting dates and years (see Chapter 5).

However, it was not until the physiology of gas exchange was applied to growth (e.g. de Wit, 1958) and dry matter changes in whole stands were examined in stands, not inside laboratories through flows of energy, carbon dioxide and water (Monteith, 1965, 1972) that substantive progress with understanding the basis of growth and yield was made. These ideas led to expressing the process of primary production as a sequence of factors linking the resource (value = 1) to yield (a very small fraction). Several of the factors were entirely of the physical environment, such as the fraction of sunlight that cloud or dust reflect or absorb. Others, such as the energy allocated to a particular structure, were largely physiological, though still responded to environment. The crucial factors linked the physical resources to the physiology: those, for example, that determine how much light is absorbed for a unit leaf area, or water for a unit root length, or how much plant material is made from a unit of energy absorbed. We return to this theme in Chapters 2 and 3.

The principles of this approach spread rapidly, and were applied in many temperate and tropical countries. Gradually, by comparison and synthesis of many studies, it became apparent that several of the factors were stable. They were not constant, but they changed only a little between seasons and fields: they were *conservative* and – as demonstrated later – they allow back-of-envelope calculations, as well as complex computer simulations, of production and yield.

Some standardization has now arisen from applying the principles of environmental physics and physiology to agriculture. Although nomenclature, symbols and units of measurement still vary, certain physical quantities have become, in effect, currencies by which to compare crops in different seasons and places. The analysis favoured in this book as a practice for tropical agronomists interprets growth and yield of different plants and crops using 'currencies' of solar radiation and water. Our approach therefore is founded on the principle that management seeks to manipulate the flows of solar radiation and water through individual plants and plant communities to a range of desired products. The factors controlling fluxes of, specifically, radiation and water are presented in Chapters 2 and 3.

1.5.5 Agronomic decisions and sequence of chapters

Following Chapters 2 and 3, the rest of this book examines the way knowledge of flux currencies can be used to inform agronomic practice. The sequence of chapters argues that choices and operations in management can be separated into two broad groups. In the first are those that govern what, when and how to sow a crop: factors that all farmers must decide around the time of sowing. They require decisions about the genotypes to be grown, the time of sowing and of other operations, and the intended configuration of the stand as influenced by the population density and species composition. Genotype, timing and configuration are examined in Chapters 4, 5 and 6, respectively. The second group of factors includes operations that are not always done, such as applying fertilizer, water-harvesting techniques and irrigation, and reducing competition from weeds. These are considered together in Chapter 7. Inevitably, several subjects are covered only briefly or have been omitted. These include the effects of pests and diseases and localized soil problems such as salinity, waterlogging and the effects of acid sulphates and peats. In practice, knowledge of the flux-currencies could help to elucidate any of these problems. A final chapter explores the wider implications of the approach and indicates where progress and deeper understanding are needed. It also looks to the exciting methodologies that might be absorbed by agronomy in the future.

2

Solar Radiation

In the absence of drought and nutrient shortages, the growth and development of crops
are ultimately controlled by the interaction
of plant systems with specific elements of
the solar spectrum. This chapter describes
the influence of solar energy on *potential*
growth and development in crops. In
Chapter 3 we consider how the availability
and demand for water often set limits on
actual crop performance.

2.1 The solar spectrum and plant processes

Although the sun radiates energy across the
whole electromagnetic spectrum, the earth's
atmosphere is transparent to all the visible
radiation but only to part of the infrared and
ultraviolet radiation that is emitted from the
solar disk. All forms of electromagnetic radiation have wave characteristics and travel at
the same speed (3×10^8 m s^{-1}) but the various types of radiation differ in *wavelength*.
Radiation can be considered in terms of discrete particles or *quanta* each of which has
an energy content that is inversely propor-

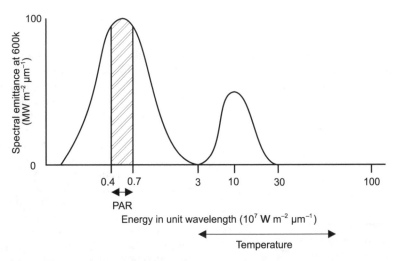

Fig. 2.1. The solar spectrum indicating the wavelength characteristics of photosynthetically active
radiation (PAR) (0.4–0.7 µm), daylength (0.66–0.73 µm) and temperature (3–100 µm) effects on the
growth and development of crops.

© CAB *International* 2002. *Principles of Tropical Agronomy*
S.N Azam-Ali and G.R. Squire

tional to its wavelength, i.e. the shorter the wavelength, the greater is the energy content. There are two major types of radiation: shortwave radiation within the waveband 0.25–3.0 µm and longwave radiation within the waveband 3–100 µm.

There are three regions of the solar spectrum that have significance for plant processes. These are: radiation within the visible range (0.4–0.7 µm), often referred to as *light*; the red (0.66 µm) to far-red (0.73 µm) range which signal changes in daylength, and longwave or terrestrial radiation which determines temperature (3–100 µm). Figure 2.1 illustrates the major regions of the solar spectrum in terms of their influence on plant processes.

2.1.1 Light and photosynthesis

Radiation within the visible range, i.e. 'light' (0.4–0.7 µm) is termed 'photosynthetically active radiation' (PAR). The energy contained within this waveband is the only radiation that can be actively used for plant growth by driving pigment-based systems in the process of photosynthesis. A quantum of light is called a photon and the process of photosynthesis is powered by photons of radiation which are absorbed by pigment molecules, such as chlorophyll, in order to convert atmospheric carbon dioxide (CO_2) to carbohydrate. The energy content of PAR varies with time and atmospheric conditions but is typically close to 50% of the total energy within the solar spectrum (Monteith and Unsworth, 1990).

Photosynthesis

Green plants must capture and use external resources, principally light, CO_2, water and nutrients, to produce dry matter via photosynthesis. By this process, plants synthesize organic compounds from inorganic materials in the presence of sunlight. The major chemical pathway in photosynthesis is the conversion of atmospheric CO_2 and water to carbohydrates and oxygen as shown in Equation 2.1:

$$CO_2 + H_2O \rightarrow CH_2O + O_2 \qquad (2.1)$$

By the input of solar radiation, two energy-poor compounds (carbon dioxide and water) are converted into two energy-rich compounds (carbohydrate and oxygen). The carbohydrates formed via photosynthesis possess more energy than the materials from which they are composed and we can consider photosynthesis as a process which reduces atmospheric CO_2 and 'converts' light energy into the chemical energy of plant tissues. So, it is not surprising that a close link exists between the photosynthetic *rate* of individual leaves and the amount of light that they absorb.

Photosynthesis reduces CO_2 to carbohydrate via two carboxylation pathways: the Calvin cycle and the Hatch-Slack pathway. In C_3 crops, the Calvin cycle predominates and the initial fixation product is a three-carbon compound. In C_4 crops, the Hatch-Slack pathway predominates and a four-carbon compound is the initial product. Here, CO_2 is refixed by the Calvin cycle and little or no carbon is lost through photorespiration (see below). The C_3 species include all the temperate crops, as well as tropical legumes, root crops and trees, whereas C_4 crops include most tropical cereals and grasses, such as maize, sorghum, millet and sugarcane.

Respiration

The high-energy compounds produced by photosynthesis are broken down (or decarboxylated) by two pathways: *photorespiration* and dark respiration. The process of photorespiration is induced in C_3 plants by the presence of oxygen. Photorespiration acts on the CO_2 initially fixed by photosynthesis and its rate is therefore closely linked to the CO_2 fixation rate. The importance of photorespiration increases with temperature, resulting in a reduction in the initial efficiency of light use of individual leaves. There is no photorespiration in C_4 plants. However, in these crops, the initial light use efficiency at low temperatures may be less than that of C_3 plants because of the energy costs involved in suppressing photorespiration.

Irrespective of their photosynthetic system, all green plants undergo the process of *dark respiration* in which atmospheric oxygen is used by plants to convert carbohydrates into CO_2 and water, with the simultaneous liberation of energy. Plants use this energy to build more complex molecules from the initial products of photosynthesis. Respiration therefore results in the loss of organic matter and oxygen from plants and a release of CO_2 to the atmosphere. Respiration is an important part of the carbon budget of crops because it is responsible for the loss of CO_2 from plant cells. It can be considered at two levels; that which occurs as a result of the *growth* of crops and that which is required for their *maintenance*. It is generally assumed that, at any given temperature, respiration continues in the light at a comparable rate to that in the dark and, moreover, that during the life of a crop, the relative contributions of the growth and maintenance components of respiration change with the age and weight of the crop. However, there have been very few measurements of respiration in the field and it is uncertain how total respiration is divided between growth and maintenance processes.

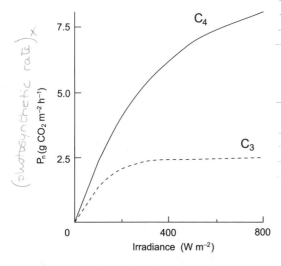

Fig. 2.2. The typical relation between photosynthetic rate and irradiance for C_3 and C_4 species. The dashed line indicates the maximum photosynthetic rate (g CO_2 m^{-2} h^{-1}) or 'saturating' irradiance (W m^{-2}) for C_3 species; C_4 species show less evidence of light saturation.

Carbon dioxide exchange

At any time, the net photosynthetic *rate* of a green plant depends on the distribution of quanta over individual elements of crop foliage, the relation between photosynthetic rate and irradiance for each element of foliage and on the corresponding rates of CO_2 loss. When light is not limiting, photosynthesis is controlled by the rate at which CO_2 from the atmosphere is reduced to carbohydrate. In very weak light, the relation for both C_3 and C_4 plant systems is almost linear because photosynthetic rate is limited almost exclusively by the absorption of light quanta. This initial slope, or 'quantum efficiency' is a measure of the amount of CO_2 absorbed per unit increase in irradiance. Typically, 8–10 photons are required to fix one molecule of CO_2 to carbohydrate. After this initial linear phase, the photosynthetic rate of C_3 species in strong light approaches a plateau at a 'saturating' irradiance with a maximum value that declines with leaf age. In contrast, C_4 species show less evidence of light saturation and, therefore, no marked plateau in photosynthetic rate at high irradiances. The apparent photosynthetic advantage of C_4 crops over C_3 crops can thus be ascribed both to the absence of photorespiration and to greater photosynthetic rates in strong light. Figure 2.2 illustrates the typical light response curves for a C_3 and a C_4 leaf.

Irrespective of species, the net photosynthetic rate, P (μmol m^{-2} s^{-1}) of any green plant is the difference between the rate at which CO_2 is reduced to carbohydrate and the rate at which carbohydrate and other compounds are oxidized to provide the energy necessary for metabolic processes (Monteith, 1981). The net flux of CO_2 into a leaf can be expressed as

$$P = \frac{(c_a - c_i)}{(r_a + r_s)} \qquad (2.2)$$

where c_a (g of CO_2 per m^3 of air) is the concentration of CO_2 in the ambient air, c_i is the concentration of CO_2 in the air spaces within the leaf, r_a (s m^{-1}) is the leaf boundary layer resistance, i.e. the resistance provided by a thin layer of still air at the leaf surface to gas transport and r_s (s m^{-1}) is the leaf stomatal resistance which depends on the distribution and degree of opening of stomatal pores on the upper and lower surfaces of the leaf. Figure 2.3a and b graphically illustrates the pathway and resistances to the entry of CO_2 into a C_3 and a C_4 leaf and the concomitant loss of water vapour and oxygen through the same apertures (see Chapter 3).

2.1.2 Daylength and morphogenesis

The earth spins around its axis once every 24 hours and at the equator all days of the year are of equal length divided into equal periods of dark and light. Strictly, the daylength at the equator is 12 hours 7 minutes rather than exactly 12 hours. (The extra 7 minutes accounts for the time taken for the upper half of the sun to disappear under the horizon at sunset and to fully appear at sunrise.) With increasing distance from the

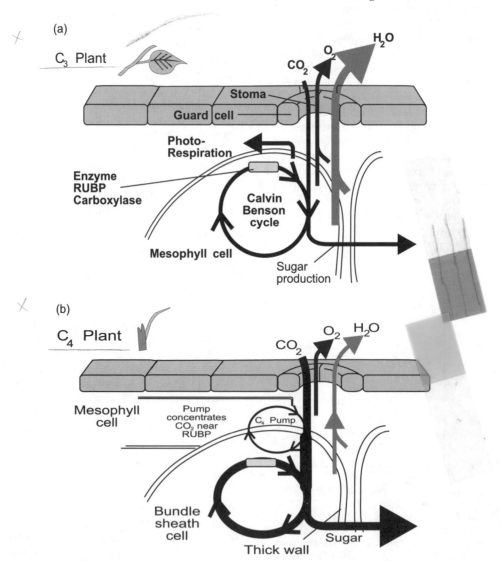

Fig. 2.3. A schematic cross-section of (a) C_3 leaf and (b) C_4 leaf showing the pathways for carbon dioxide entry and water vapour loss.

equator, the difference between the shortest and longest day of the year increases. At low latitudes the increase is about 7 minutes per degree but at higher latitudes the difference is larger. For example, between 50° and 60° it is about 28 minutes per degree of latitude.

Changes in visible radiation in the red (0.66 µm) to far-red (0.73 µm) range can have profound effects on many organisms. In terms of plant behaviour, changes in red : far-red can alter the pattern of growth and development of some plant species. This phenomenon, known as photomorphogenesis, is responsible for the flowering behaviour of many species. More precisely, it is changes in the ratio of red to far-red radiation that influence the reproductive behaviour of plants. Above a solar elevation of about 10° the ratio remains constant at about 1:1. However, at sunrise and sunset the ratio changes rapidly. Periodic changes in the ratio from winter to summer are responsible for *daylength* effects in plant species. The perception of changes in daylength can be considered as an 'informational' resource in contrast with PAR which can be considered as a 'growth' resource.

In some crops, an information signal is required for them to switch from vegetative to reproductive activity. When this signal is provided by changes in daylength the phenomenon is known as *photoperiodism*. Thus, crop species can be defined as *photoperiod-sensitive* or *photoperiod-insensitive*. Species that are sensitive to changes in daylength can be further sub-divided into those which flower in response to short days ('short-day' plants) and those which respond to long days ('long-day' plants). Species that are unaffected by daylength are known as 'day-neutral' plants. Within species that are predominantly short-day or long-day, there exist varieties that may also be day neutral.

2.1.3 Temperature and plant development

Both the longer wavelengths of earthward solar radiation and long-wave terrestrial radiation, i.e. reflected radiation from the ground, are insufficiently energetic to drive pigment systems such as photosynthesis. However, the balance between incoming and outgoing radiation is primarily responsible for the temperature of plants and soils, which strongly controls the *development* rates of plants.

Germination

A good example of the dominance of temperature on developmental events is the germination of seeds. For germination to occur, seeds must be viable, non-dormant and adequately supplied with water and oxygen. For most of the major crop plants so far examined, the initiation of germination in each species requires a minimum base temperature (T_b). Thereafter, germination rate increases linearly with temperature to an optimum value (T_o), above which the rate declines linearly to a maximum temperature (T_m). In some species, this optimum temperature may cover a range of several degrees rather than a single temperature. The typical relationship between germination and temperature is illustrated in Fig. 2.4. When the temperature, T, is above T_b and below T_o the relation between T and $1/t$ is given by

$$\frac{1}{t} = \frac{(T - T_b)}{\theta_1} \qquad (2.3)$$

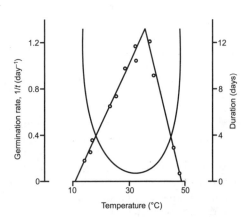

Fig. 2.4. The typical relation between germination rate (expressed as the reciprocal of time, *t*) and temperature for viable seeds (open symbols). The solid line without data points indicates the same germination sequence expressed in days. (Adapted from Squire, 1990.)

and when T is above T_0 and below T_m the relation is given by

$$\frac{1}{t} = \frac{(T_m - T)}{\theta_2} \qquad (2.4)$$

The constants θ_1 and θ_2 represent the thermal duration for the developmental process of germination to occur.

Thermal time

Because the quantity θ is the product of temperature and time it is measured in degree-days ($^\circ C^d$) and is called *thermal time*. In practice, the value of T_b is usually similar for different developmental processes within a species and thus the thermal time for any process can be defined as,

$$\theta = \sum_{i=1}^{i=n} (T - T_b) \qquad (2.5)$$

where n is the number of days experienced by a plant at a mean daily temperature, T, above T_b. Thus, on a day when the temperature remains below T_0, the thermal time accumulated by a crop is simply the difference between the mean daily temperature and the value of T_b.

Thermal time is an extremely useful concept because it allows the development of crops at different locations and in different seasons to be compared. This is particularly true for determinate crops, which have a clearly defined point at which seeds are physiologically mature. Often, the same crop variety is grown at a number of sites and so requires different amounts of *chronological* time, but similar amounts of *thermal* time, to reach maturity. Thus, although the chronological duration for any particular developmental process is shortest at the optimum temperature and lengthens at temperatures below (or above) this value, the thermal duration for the same process remains similar at each location.

Where the switch to reproductive growth does not require a specific daylength (i.e. day-neutral or daylength insensitive plant types), the date of flowering in tropical crops can be estimated from a knowledge of thermal time.

The concept of thermal time remains valid over a wide range of environmental conditions and is largely unaffected by changes in solar radiation. However, thermal time is often sensitive to water deficits.

Developmental windows for canopy expansion

The prevailing temperature and photoperiod set the potential duration of each developmental phase of a crop. The amount of growth that occurs within each developmental phase is set by the amount of solar radiation intercepted by plant canopies.

The rate at which leaves are produced and expand early in the season determines the amount of radiation that a plant can intercept. Generally, temperature and photoperiod determine the number of leaves on a plant, while temperature, water status and nutrients dominate the extension rates of individual leaves. So, to maximize the size of its intercepting surface, a plant must expand individual leaves as rapidly as possible within the developmental period set by thermal time.

To expand their leaves and synthesize organic compounds, plants must take up inorganic nutrients from the soil. Although many elements are required for optimal plant growth, most of these are required in such small amounts that the supply from the seed or from natural sources is often adequate (Van Keulen and Wolf, 1989). However, the macro-elements nitrogen, phosphorus and potassium are usually needed in such large quantities that they have to be provided as fertilizers to supplement the natural supply. This is especially true where management practices aim at very high yields. Nitrogen is the element that is needed in the largest quantities, and major deficiencies in plant nitrogen can directly affect transpiration, carbon assimilation and partitioning to various organs. However, it is the crucial role of nitrogen in the expansion of leaves that perhaps most strongly determines the productivity of a plant.

Because nitrogen is needed in the plant to synthesize new tissue, the total demand for nitrogen increases with increasing plant weight. However, as more structural carbohydrate is accumulated, the ratio of nitrogen to total biomass in each of the plant parts falls, even when nitrogen is available in surplus (Van Keulen and Wolf, 1989). The timing and method of application and rate of uptake by plant roots is critical to the proportion of applied nitrogen that is utilized by individual plants (see Chapter 7).

2.2 The fate of radiation in crop systems

In this discussion, we have seen the critical role that solar radiation plays in the behaviour of crops. It is therefore not surprising that most management decisions, e.g. sowing date, planting density, fertilizer application and irrigation, aim to modify the shape and extent of crop canopies and the way in which they intercept radiation. We now consider the efficiency with which crops convert solar energy into the chemical energy of plant products.

The productivity of a plant system can be defined in terms of the efficiencies with which the solar resource is captured and converted into biomass (Monteith, 1972; Azam-Ali *et al.*, 1994) i.e. the overall efficiency, ε, of the system can be described as

$$\varepsilon = \varepsilon_g \varepsilon_a \varepsilon_p \varepsilon_i \varepsilon_s \varepsilon_q \varepsilon_r \varepsilon_H \qquad (2.6)$$

where, ε_g is a geometrical ratio, i.e. the ratio of solar energy flux to solar constant. This figure represents the proportion of energy received at any particular latitude relative to the irradiance (typically 1373 W m^{-2} or 1.37 kJ m^{-2} s^{-1}) received on a perpendicular surface located at the mean distance of the earth from the sun. The value ε_a represents the atmospheric transparency, i.e. the proportion of radiation that passes through the atmosphere, ε_p is the spectral ratio, i.e. PAR as a fraction of total radiation, ε_i is the interception efficiency of foliage, ε_s is the conversion efficiency of intercepted radiation, ε_q is the photochemical efficiency, i.e. the energy value of carbohydrates, proteins and lipids, ε_r is the respiration efficiency and ε_H

is the yield efficiency, i.e. the proportion of crop weight that is reproductive or economic yield.

2.2.1 ε_g, ε_a and ε_p

These all influence the amount of photosynthetically useful radiation that is received above a field crop. The value of ε_g is determined by the geometry of the earth's surface in relation to the sun, which in turn depends on latitude and season. The annual average value of ε_g decreases from about 0.3 in the tropics to 0.2 in temperate latitudes (Monteith, 1972).

The value of solar radiation that passes through the atmosphere (ε_a) depends on how much is absorbed and scattered by gases, clouds and aerosols containing dust, smoke and biological material, such as spores and insects. When solar radiation collides with small gas molecules of the atmosphere, the resulting diffusion is known as 'Rayleigh scattering'. This diffusion is inversely proportional to the fourth power of the wavelength – in other words, short wavelengths are more scattered than long wavelengths. The mean global transmission caused by Rayleigh scattering is about 0.91 for direct-beam solar radiation (Bonhomme, 1993). Where diffusion occurs due to aerosols in the atmosphere, the degree of scattering depends on the particle sizes within the aerosols. Because water vapour contains droplets that are larger than the wavelength of the solar radiation, clouds scatter the radiation uniformly for all wavelengths. The result of these various processes is that, where the turbidity of the atmosphere is small (i.e. the atmosphere is transparent), the diffused radiation contains a greater proportion of short wavelengths and, because of Rayleigh scattering, the sky is 'blue.' In contrast, when the sky is cloudy, the diffused radiation has a spectral composition that is similar to direct beam radiation. Bonhomme (1993) provides a calculation of mean global solar radiation, R_s, that is composed of a direct-beam (R_b) and a diffuse (R_d) component. His calculated average of 11 MJ m^{-2} for total daily extraterrestrial radiation for the whole of the earth's surface is distributed as follows (in MJ m^{-2}):

0.9 scattered by molecules and aerosols back to space

0.7 absorbed by the gaseous components of the atmosphere

2.9 scattered by clouds towards space

1.2 absorbed by clouds

5.3 reaches the ground, i.e. R_s which is partitioned into $R_b = 3.3$ and $R_d = 2.0$.

The ratio of PAR to direct-beam solar radiation is approximately 0.45 (Monteith, 1965) and that for diffuse solar radiation is about 0.60 (Sinclair and Muchow, 1999). When the direct and diffuse components are combined, Monteith (1972) estimates that a value of 0.50 for ε_p is appropriate for the tropics.

Although they each have important influences on the availability of solar radiation to crop canopies across the globe, and therefore the rate and duration of crop growth, in practice there is nothing that can be done by an individual grower to alter ε_g, ε_a or ε_p at any particular location. In practical terms, there are only two processes that can directly influence the performance of a field crop in relation to solar energy. Essentially, the behaviour of solar radiation in crop stands can be described in terms of their *capture* and *conversion* systems.

2.2.2 Capture of resource

For a crop to produce dry matter, its leaves must intercept radiation and absorb CO_2. The size and duration of crop foliage determine the rate and duration of dry matter accumulation. The size of the intercepting surface depends on the green leaf area index (L) of a crop which can be expressed as the product of the number of plants per unit of ground area, the number of leaves per plant and the mean area of leaves per plant.

The amount of light that penetrates the canopy and strikes the ground depends both on L and on the angular arrangement of individual leaves. To describe the pattern of

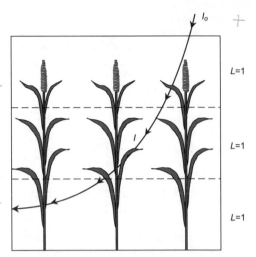

Fig. 2.5. The exponential decay of radiation through a crop stand. In this illustration, the canopy is divided into three layers of uniform leaf area index ($L = 1$). The radiation, I, received at any level in the canopy is expressed as $I/I_0 = e^{-kL}$, where I_0 is irradiance and k is the extinction coefficient for the species.

light penetration through a crop canopy, it is convenient to imagine a crop as consisting of a number of horizontal layers each of $L = 1$ (see Fig. 2.5). If radiation is measured at a number of levels down the crop profile, each corresponding to unit L, then the measured irradiance at any level is a function of the angular arrangement of leaves above that level. The relationship for the extinction of light down a crop canopy is often described by the Monsi–Saeki (1953) equation:

$$\frac{I}{I_0} = e^{-kL} \qquad (2.7)$$

where I_0 is the irradiance above the crop canopy, I is the irradiance at a level within the canopy below a leaf area index of L and k is an extinction coefficient for radiation. However, it should be noted that the Monsi–Saeki equation assumes that the canopy is a homogeneous medium whose leaves are randomly distributed (such that there is no effect of row structure or clumping on the pattern of light transmission through the canopy).

In these (ideal) circumstances, light transmission obeys Beer's law of exponential decay. Strictly, for attenuation to be

Table 2.1. The range of some reported values for the extinction coefficient (*k*) for ten major crops of the world. (From Azam-Ali *et al.*, 1994.)

Wheat	0.40–0.70
Rice	0.43–0.86
Maize	0.56–0.78
Barley	0.48–0.69
Sorghum	0.43–0.60
Millet	0.30–0.59
Cassava	0.58
Soybean	0.45–0.96
Groundnut	0.40–0.66

exponential, the leaves should be black, i.e. opaque to radiation. However, Goudriaan (1977) has shown Beer's law to be a good approximation in many real canopies and, in these circumstances, a plot of $\ln(I/I_0)$ against *L* gives a straight line with a gradient *k* which depends on the architecture of the crop. Table 2.1 presents some reported ranges in the value of *k* for a number of major crops. Crops with narrow, erect leaves tend to have lower values of *k* than crops with more horizontally displayed leaf arrangements.

If the radiation transmitted through the canopy follows the Monsi–Saeki equation and the extinction coefficient is known, the fraction (*f*) of radiation intercepted by a crop can be calculated from a knowledge of *L*, i.e.

$$f = 1 - \exp(-kL). \qquad (2.8)$$

In many tropical systems, crops rarely, if ever, cover the ground completely. This can be because crops are deliberately sown in distinct clumps or rows to optimize the use of available water rather than light or because the hostile soil environment prevents uniform and complete crop establishment. In these circumstances, the Beer's law analogy of randomly distributed leaves and the corresponding Monsi–Saeki equation fails. Figure 2.6 illustrates the type of radiation distribution in stands where crops are arranged in distinct rows compared with a more random arrangement. We return to the

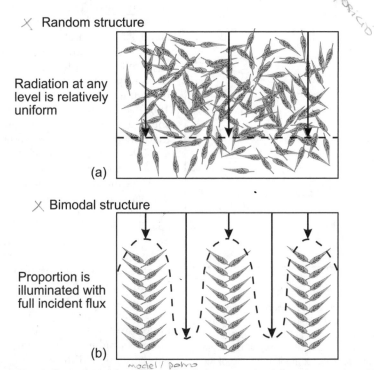

Fig. 2.6. A diagrammatic illustration of the pattern of radiation transmitted through a canopy with (a) randomly distributed leaves (b) bimodal structure.

implications of non-random canopy structure in Chapters 7 and 8.

The role of nitrogen in light capture

Competition for light between individual
plants within a crop canopy reduces the
nitrogen concentration (N%) of individual
plants (Lemaire and Gastal, 1997). Several
authors have shown that the reduction of
N% through the vertical profile of several
species follows the shape of the attenuation
profile for light (Field, 1983; Charles-
Edwards et al., 1987; Pons et al., 1993).
Figure 2.7 shows that when leaf N content
is expressed on a leaf area basis, a close relation is often found between the irradiance
(I_z) at any depth within the canopy and the
N% per unit leaf area (N_z) (Hirose et al.,
1988; Lemaire et al., 1991; Lemaire and
Gastal, 1997). There appear to be two
aspects to the decline in N% with depth in
the canopy. First, there is the decrease in
N% with light attenuation as shown in Fig.
2.7 to a light extinction level corresponding
to the compensation point for net photosynthesis. Second, there is a rapid decrease

beyond this point, when the carbon balance
of each leaf is negative and protein maintenance becomes impossible, leading to leaf
senescence and abscission. Measurements
of N content per unit leaf area on yellow
leaves collected immediately after abscission indicate that up to 80% of the original
leaf N content is recycled to the remainder
of the plant (Lemaire et al., 1991). Any nitrogen remaining in the dead leaf is assumed
to be structural N, whilst the remobilized
fraction corresponds to the metabolic pool
(Lemaire and Gastal, 1997).

2.2.3 Conversion of resource

Conversion efficiency of solar radiation (ε_s)

To calculate the productivity of a stand, the
photosynthetic rate of individual leaves
must be integrated with respect to space and
time. The rate of leaf photosynthesis is limited by the rate at which it can assimilate
CO_2 from the atmosphere and reduce it to
carbohydrate. The theoretical minimum
energy required to reduce a CO_2 molecule is
about 9.5 quanta of PAR (Penning de Vries
et al., 1989). However, the lowest value
observed in C_3 plants, in the absence of
photorespiration, is about 13 quanta per
molecule (Farquhar and von Caemmerer,
1982). (The difference is largely because
some light is absorbed by non-photosynthetic pigments.) Nevertheless, leaf photosynthesis is initially limited by the amount
of radiant energy incident on the leaf surface. As the incident energy increases, the
concentration of CO_2 in the air around the
leaves limits the rate of photosynthesis. The
spatial distribution of leaves within a crop
canopy and the extent to which individual
leaves are shaded, therefore, markedly affect
whether canopy photosynthesis is light or
CO_2 limited.

For uniform crop stands, the spatial
integration of photosynthesis can again be
approximated by the Monsi–Saeki model,
and the irradiance that each leaf intercepts
can be modelled in terms of the mean orientation of leaves in relation to incident
radiation. Individual leaf photosynthesis

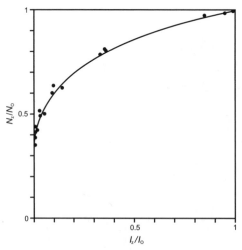

Fig. 2.7. The relationship between relative
irradiance at depth z within a canopy (I_z/I_0) and
the relative nitrogen content per unit leaf area
(N_z/N_0) for a lucerne stand. I_0 and N_0 represent
the irradiance and leaf content at the top of the
canopy. (Taken from Lemaire et al., 1991;
Lemaire and Gastal, 1997.)

therefore occurs at a rate which depends on irradiance and the relation between photosynthetic rate and irradiance for each element of foliage.

The temporal integration of photosynthesis requires a knowledge of crop behaviour over the whole growing season. Experimentally, a short-cut can be taken by expressing dry matter accumulation (measured by destructive sampling) in relation to intercepted radiation obtained by sequential measurements of radiation above and below the crop (Sceicz, 1974). Therefore, the rate of dry matter production per unit time, $\delta W/\delta t$, can be expressed as

$$\frac{\delta W}{\delta t} = \varepsilon_s fI_0 \qquad (2.9)$$

where, fI_0 (MJ m^{-2}) is the amount of light intercepted by the crop (i.e. the product of the fractional interception, f, and the daily total of incident radiation, I_0) and ε_s (g MJ^{-1}) is the efficiency with which the crop uses that light to produce dry matter.

The total biomass accumulated over a growing season can now be expressed as

$$W = \varepsilon_s \int fI_0 dt. \qquad (2.10)$$

The relation between dry matter and accumulated intercepted radiation as described in Equation 2.10 remains stable in many crops, particularly during vegetative growth. As long as its value is conservative, a knowledge of ε_s therefore allows us to describe the photosynthetic productivity of a crop in terms of captured solar radiation. However, there are many circumstances where ε_s is lower than expected for a given species or appears to decrease during the life of a crop. There are at least three general types of departure from the linear relation between total dry matter and accumulated intercepted radiation. The slope of the relation may:

(a) *Decrease and eventually level off during reproductive growth*. This well-documented phenomenon occurs in many determinate crops, such as cereals, and may result from a number of causes. For example, Gallagher and Biscoe (1978) found that the value of ε_s for PAR in barley and wheat crops fell by

30% after ear emergence. They ascribed this decline to both a reduced photosynthetic rate by ageing leaves and an increase in respiration per unit of assimilation after anthesis. However, in many species the production of new leaves delays the impact that senescing older leaves have on the value of ε_s until leaf production stops. Because temperature determines the duration of the reproductive phase in determinate crops which are not limited by stresses, such as drought, the proportion of time that crops intercept radiation when ε_s is not constant will depend on seasonal weather.

(b) *Break down during the season*. This may occur through drought, pests and/or diseases. In each of these cases, ε_s may decline at any time after stress is imposed on the plant system. Drought reduces ε_s by decreasing photosynthetic capacity and leaf conductance. As a result, the rate of dry matter production may decrease or eventually stop even though the crop canopy continues to intercept radiation (Steduto and Hsiao, 1994).

Pests and diseases can have a complex influence on ε_s. They may directly affect canopy structure by removing active leaf area so that a greater proportion of the intercepting surfaces of the crop are composed of non-photosynthetic elements, e.g. stems, flowers, grain. Alternatively, pests and diseases can affect canopy function and photosynthesis by disrupting leaf tissue and/or by removing assimilates that have already been fixed by the crop.

(c) *Be reduced in gradient throughout the season*. The value of ε_s is sometimes less than that expected for a particular species throughout the season. This may occur as a result of nutrient stress or because of a large saturation deficit, D (kPa), of the air surrounding the canopy. In both cases, photosynthetic rates may be reduced throughout the life of the crop resulting in a lower slope for the relation between dry matter and intercepted radiation.

There have been few attempts to quantify the effect of nutrient supply on the value of ε_s. However, Muchow and Davis (1988) found a reduction in ε_s for maize and sorghum growing in nutrient poor soils in

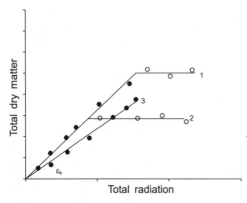

Fig. 2.8. Schematic examples of how the relation between crop dry weight and accumulated intercepted radiation can break down: (1) late in the season during reproductive growth; (2) at any stage in the season when stress limits dry matter production, e.g. drought, and (3) throughout the season when stress limits dry matter production, e.g. nutrient shortage.

Australia, and Muchow (1992) showed a similar effect in kenaf where the value of ε_s for a crop supplied with 240 kg ha^{-1} declined from 1.20 g MJ^{-1} to 0.94 g MJ^{-1} in a crop which was not supplied with nitrogen. Squire (1990) described a similar reduction in ε_s in relation to the leaf potassium content of oil palm fronds growing in Malaysia. He also noted that for the sorghum, maize and oil palm examples given above, the decline in ε_s was greater than the reduction in the fraction (f) of radiation intercepted by the crops and that measured changes in f and ε_s are seldom consistent between sites, even for one genotype.

The sensitivity of ε_s to saturation deficit, even under non-stressed conditions has previously been demonstrated for field crops of sorghum and maize (Stockle and Kiniry, 1990) and for sorghum growing in controlled-environment glasshouses (Hamdi *et al.*, 1987). Stockle and Kiniry (1990) reported that the average daily mean value of D accounted for 76% of the variability in ε_s for sorghum and 50% of the variability in ε_s for maize for crops that were reported as having no water, nutrient or low temperature stresses. The three types of departure from the linear relation between

dry matter and radiation are illustrated in Fig. 2.8. Table 2.2 summarizes some of the reported ranges in the value of ε_s for various crops under non-limiting and apparently limiting conditions. Sinclair and Muchow (1999) give a more complete review of radiation conversion efficiency.

Photochemical efficiency (ε_q)

The process of plant growth consists of the conversion of glucose to other organic compounds, the translocation of the glucose to the site of growth and, in legumes, the cost of nitrogen fixation. In terms of relative energy contents, plant tissue is composed of varying proportions of five distinct biochemical groups; nitrogenous compounds (especially proteins), carbohydrates (cellulose, hemicellulose, starch), lipids (fats, fatty acids, oils), lignin and organic acids. The weight ratio of these biochemical products to substrate varies between 0.35 and 1.0 g g^{-1} (Penning de Vries *et al.*, 1989). This is because the synthesis of products that are

Table 2.2. Values of the conversion coefficient (g MJ^{-1}) for intercepted radiation (total solar) in some major crops of the world. (From Azam-Ali *et al.*, 1994.)

Crop	ε_s	$\varepsilon_s{'}$
C$_4$ species		
Maize	2.43	1.30
Sorghum	2.69	1.20
Millet	2.62	0.57
C$_3$ species		
Wheat	1.50	0.80
Rice	2.05	1.00
Barley	1.20	1.10
Potato	1.84	1.00
Cassava	–	0.90
Sweet potato	1.55	–
Soybean	1.30	0.23
Groundnut	1.47	0.47

The maximum value (ε_s) was derived from experiments under apparently optimal conditions. The constrained value ($\varepsilon_s{'}$) was derived from experiments where shortage of water and/or nutrients appear to have reduced the conversion efficiency but individual plants have continued to grow.

Table 2.3. Some characteristics of the five major groups of plant components and minerals. (From Penning de Vries *et al.*, 1989.)

	Heat of combustion (kJ g^{-1})	Nitrogen content (g g^{-1})	Carbon content (g g^{-1})
Carbohydrates	17.3	0.00	0.45
Proteins	22.7	0.15	0.53
Lipids	37.7	0.00	0.77
Lignins	29.9	0.00	0.69
Organic acids	13.9	0.00	0.38
Minerals (K, Ca, P, S)	0.0	0.00	0.00

rich in lipids and proteins requires considerably more energy per unit dry weight than does the synthesis of carbohydrates. Table 2.3 summarizes the energy content of the different plant compounds in relation to their biochemical composition. From this, it can be seen that the lipid fraction of seeds contains about 2.19 times as much energy as an equivalent weight of carbohydrate. Therefore, calculations of ε_s based on dry weight need to be adjusted to account for the energy equivalents of the lipid and protein components. Clearly, the unadjusted value of ε_s must decrease during the grain filling of crops which produce seeds that are high in oil or protein (Muchow *et al.*, 1982; Kiniry *et al.*, 1989).

es during the season because of an increase in maintenance respiration (McCree and Silsbury, 1978). In temperate regions, the seasonal increase in temperature should also cause a reduction in ε_s. However, Squire (1990) has summarized evidence to show that ε_s varies little with plant mass whilst Kiniry *et al.* (1989), working with five species in the USA, showed no effects of changing temperatures on the value of ε_s. They concluded that energy loss due to respiration appeared to be proportional to plant growth. Recent evidence (Albrizio, 2001) demonstrated that the rate of dark respiration in chickpea, wheat and sorghum stands was linearly related to assimilation rate irrespective of plant age or temperature.

Respiration efficiency (ε_r)

One method of quantifying the significance of respiration to the total carbon balance of a crop is to express *growth efficiency* as the ratio of the increase in biomass to the sum of the increase in biomass and respiration over the same period. Amthor (1989) estimated that the growth efficiency of several field crops ranged between about 0.5 and 0.6, i.e. about half the carbon assimilated in photosynthesis is eventually lost in respiration.

Accurate measurements of respiration are difficult because the photosynthetic rate, temperature, size and age of a crop vary throughout its life and affect the relative amounts of photosynthesis and respiration. Therefore, the exact influence of respiration on ε_s is extremely difficult to assess. Nevertheless, in theory the value of ε_s should decrease as crop dry weight increas-

Yield efficiency (ε_H)

For many determinate crops, the reproductive weight of individual plants, g, is closely related to the total dry weight, w, of each plant above a minimum weight, w_0, i.e. the minimum amount of vegetative 'infrastructure' necessary before reproductive growth can commence. Provided that total plant weight is much greater than w_0, as it is for many well tended crops, the ratio of reproductive (or economic) yield to total dry weight, i.e. the harvest index, H, remains conservative. Therefore,

$$g = H(w - w_0) \qquad (2.11)$$

In these circumstances, the total reproductive yield, G, per unit ground area is given by

$$G = P_g = H(W - Pw_0) \qquad (2.12)$$

Table 2.4. Reported values of harvest indices[a] for some major crops of the world. (From Azam-Ali *et al.*, 1994.)

Crop	H	H'
Wheat[b]	0.50	0.22
Rice	0.50	0.25
Maize	0.45	0.30
Barley	0.52	0.35
Sorghum	0.40	0.20
Oat	0.50	0.23
Millet	0.45	0.20
Rye	0.29	–
Potato[c]	0.65	0.55
Cassava[d]	0.77	0.30
Sweet potato[e]	0.74	0.50
Soybean	0.59	0.25
Groundnut	0.48	0.16
Sugarcane[f]	0.30	0.20

[a] The maximum harvest index (*H*) is presented for crops growing under apparently optimal conditions. The comparable constrained harvest index (*H'*) is shown for crops whose growth appears to have been limited by adverse environmental factors; [b] refers to bread and durum wheat; [c] fresh tuber at 32.5% dry weight; [d] fresh tuber at 35.0% dry weight; [e] fresh tuber at 30.0% dry weight; [f] sugar at 10–12% of fresh cane.

where *P* is the plant population and *W* is the total dry weight per unit area of ground.

However, the allocation of assimilate to the reproductive or economic components is not always conservative (see Table 2.4) and estimates of yield based on such an assumption may be very wrong. Again, this is particularly the case for crops which are grown in marginal areas, often on a stored water profile. Here, the vegetative phase may continue more-or-less as normal whilst there is adequate water but drought will become increasingly important during grain filling. This will lead to premature senescence of leaves and a reduction in crop photosynthetic potential. The net effect will be a crop with a reasonable vegetative growth but poor final yield, i.e. a lower value of *H*.

2.3 Conclusions

In this chapter, we have seen how a knowledge of intercepted radiation and the conversion efficiency of that radiation provide a useful basis to analyse crop growth and yield. The importance of interception and conversion of solar radiation is obvious since it is light that determines the potential rate of photosynthesis. Of course, CO_2 is an essential substrate for photosynthesis but, in practice, this cannot be manipulated in the field and changes little between sites or seasons. The main drawback to using the capture and conversion of solar radiation as a simple and apparently effective scheme for calculating crop growth is that for much if not all of the life of many tropical crops, the remaining two resources, water and nutrients, constrain growth below the potential set by the interception and conversion of light.

3

Water

Chapter 2 considered crop growth in relation to the capture and conversion of solar radiation which allows us to quantify production efficiency in terms of the dry matter accumulated by crops relative to the amount of solar energy intercepted by their foliage. It was shown that for many crops, the link between dry matter and intercepted radiation is linear, particularly during vegetative growth, and a 'conversion coefficient for intercepted radiation' (ε_s; g MJ^{-1}) was introduced to quantify this relation. Plants do not directly 'convert' radiation into matter but use photosynthetically active radiation to drive the process of photosynthesis through which CO_2 is converted into carbohydrate. The CO_2 required for photosynthesis has to be obtained from the atmosphere through stomata on the leaf surfaces. For a crop, it is both the size and frequency of stomata and the area of the leaf surfaces that determine how much CO_2 is absorbed. However, an inevitable consequence of stomatal uptake of CO_2 is that water is lost through the same apertures. This chapter demonstrates that, for any crop, there is a functional link between the amount of CO_2 absorbed and the amount of water lost through transpiration. Therefore, instead of using ε_s as a basis for calculating crop productivity, we can consider the efficiency of crop production in terms of the dry matter accumulated relative to the amount of water transpired.

When considering the link between crop productivity and transpiration, it is tempting to use rainfall as a guide. However, seasonal rainfall and transpiration are not synonymous and it is essential to emphasize that it is *seasonal transpiration* that is associated with crop productivity. Before examining the link between growth and transpiration, we need to examine first how soil and atmospheric factors determine the proportion of rainfall that is made available for use by crops and how, once extracted, this soil water is taken up and used by plants.

3.1 The hydrological cycle

The global hydrological cycle has four components – the soil, groundwaters, streams and seas (Briggs and Courtney, 1989). These are linked by various transfer processes such as drainage, throughflow, runoff and seepage. Differences between the various components of the hydrological cycle are crucial in determining the type of vegetation and its net productivity within any region. In this context, the critical link within the hydrological cycle is the balance between the supply of water, mainly via rainfall, and the demand for water through evaporation from plant and soil surfaces.

In terms of crop production, rather than the whole global hydrological cycle, it is appropriate to consider the *water balance* of the field site in which a particular crop is to be grown (see Fig. 3.1). The water balance at any location can be expressed as

© CAB *International* 2002. *Principles of Tropical Agronomy*
S.N Azam-Ali and G.R. Squire

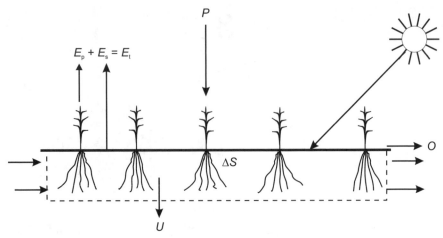

Fig. 3.1. The hydrological cycle of an arable field. Various factors are included to show the components that control evapotranspiration (E_t) and its components, i.e. evaporation directly from the soil surface (E_s) and transpiration (E_p). Other terms include precipitation (P), runoff (O), deep drainage (U) and soil moisture storage (ΔS).

$$P - O - U - E_t - \Delta S = 0 \qquad (3.1)$$

where P is precipitation, O is the surface runoff, U is the flow of water from the soil surface to underlying layers (i.e. deep drainage), E_t is the total evaporation from plant (E_p) and soil surfaces (E_s) and ΔS is the net change in soil water storage. It is this store of soil water that plant roots must access to meet the transpirational requirements of the leaf canopy. However, even an accurate water balance that provides an exact knowledge of how much water is *present* within a soil does not tell us how much of that water is *available* for uptake by plant roots. For this, we need to consider concepts of soil water energy.

3.2 Soil water energy

To understand how water behaves within soils we need to consider the concept of *free energy*. The term 'free energy' is used to describe the energy status of water and represents the summation of all other forms of energy that are available to do *work*. Water, like all substances, has a tendency to move from a state of higher to one of lower free

energy. In this way, water will tend to move from a wet region of soil (high energy state) to a drier region of soil (low energy state). It is the difference in the energy status of water at various points in the soil that influences the direction of water movement.

Consider a given volume, V, of soil which consists of a volume, V_s, of solids (inorganic plus organic particles), V_w, water (liquid plus vapour) and V_a, air, so that:

$$V = V_s + V_w + V_a. \qquad (3.2)$$

The volume $(V - V_s)$ is the total volume of pore space and the ratio

$$e = \frac{(V - V_s)}{V} \qquad (3.3)$$

is defined as the *porosity* of the soil, i.e. the fraction of any total volume of porous material occupied by pores.

The *water content* of soil can be expressed on a volume basis as

$$\phi_s = \frac{V_w}{(V - V_s)} \qquad (3.4)$$

where ϕ_s is known as the *saturation ratio* of

the soil expressed as a percentage. Water content can also be expressed on a mass basis as

$$w = \frac{m_w}{m_s} \qquad (3.5)$$

where m_w and m_s are the masses of soil water and dry solids respectively. Water content expressed on a mass basis can be converted to a volume basis from a knowledge of bulk density (ρ_b) (the weight per unit volume of dry soil), that is,

$$\phi = \frac{w\rho_b}{\rho} \qquad (3.6)$$

where ρ is the density of water. The term ϕ actually expresses the equivalent depth of water that exists per unit depth of soil.

Because it quantifies how much water there is in a 'tank' of soil, a knowledge of ϕ might seem a sufficient basis to determine how much water is available for use by plants. However, the retention, movement and uptake of water from the soil and its loss to the atmosphere are all processes that require energy. As a consequence of these energy-dependent processes, there are differences in the *availability* of water in different soil types, even when they contain the same *amount* of water. This is because the availability of water depends on the free energy status of the water in the soil at the time of extraction by plant roots.

There are three important forces that affect the free energy of soil water. The adhesive attraction of soil solids for water within the soil matrix provides a *matric* force. Similarly, the cohesive attraction of ions and other solutes for water results in *osmotic* forces which tend to reduce the free energy of water in the soil solution. Finally, *gravitational* forces cause water to move downwards because the free energy of water at some point in the soil profile is greater than at a point lower in the profile. The difference between the free energy of soil water and that of pure water in a standard reference state is known as *soil water potential*, ψ_s. From the above it can be seen that the total value of ψ_s has three main components,

i.e. gravitational potential, ψ_g, matric potential, ψ_m, and osmotic potential, ψ_o such that

$$\psi_s = \psi_m + \psi_o + \psi_g. \qquad (3.7)$$

GRAVITATIONAL POTENTIAL. Gravity plays a crucial role in drainage of water from upper to lower soil levels with a resultant loss of free energy. The gravitational potential of water can be calculated from

$$\psi_g = g_h \qquad (3.8)$$

where g is acceleration due to gravity and h is height of the soil water above a reference level.

MATRIC POTENTIAL. This is the result of two surface phenomena, adhesion and capillarity, and expresses the attraction of soil solids and their exchangeable ions for water. Together, adhesion and capillarity reduce the free energy of soil water when compared with pure unabsorbed water. Therefore, ψ_m is always negative.

OSMOTIC POTENTIAL. Unlike ψ_m, osmotic potential ψ_o has little influence on the mass movement of water in soils but has a major influence on the uptake of water by plant roots. In soils that have high concentrations of soluble salts, ψ_o may be greater (i.e. the water is more 'concentrated') in the soil solution than in the root cells. As a result, the uptake of water by the plant roots is constrained.

UNITS OF MEASUREMENT. Water potential is commonly expressed in terms of the standard atmospheric pressure at sea level, i.e. 760 mmHg or 1020 cm of water. One bar is equivalent to a standard atmosphere and the SI unit megapascal (MPa) is equal to 10 bar.

3.2.1 Soil water potential and content

There is an inverse relation between the amount of water in a soil and the tenacity with which it is held, i.e. the less water there is in a soil the harder it is for roots to extract it. Two of the many factors that affect the

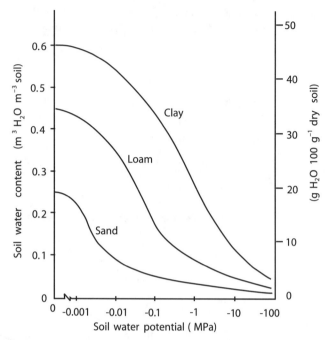

Fig. 3.2. The relation between soil moisture content and soil water potential for three representative soils.

relation between soil water content ϕ and soil water potential ψ_s are the *texture* and *structure* of a soil. For example, a clay soil holds water much more tightly at a given soil water content than does a sandy soil. A large proportion of the water held by clay soils is held so tightly that it cannot be removed by growing plants (Brady and Weil, 1999).

The structure of a soil can also influence the soil moisture–energy relationship. For example, a compacted soil is able to hold less water than a friable soil which contains a greater proportion of pore spaces. The compacted soil may also have a greater proportion of small pore spaces which tend to hold water more tightly than do larger pores. Again, the result is that at a given soil moisture content less water is available to plants in a compacted soil than in a well-granulated soil. Figure 3.2 shows the relation between ϕ and ψ_s for three soil types.

3.2.2 Water movement in soils

Water moves through the soil either as a liquid or as a vapour. The flow of liquid water can either be through saturated soil, where the soil pores are completely filled with water, or through unsaturated soil, where soil pores are only partly filled with water. In either case, the flow of water is along a gradient of matric potential from a zone with a higher value of ψ_m to one at a lower value of ψ_m. The relative influence of gravitational and osmotic forces on water movement will depend on the saturated or unsaturated state of the soil.

SATURATED FLOW. Following heavy rain or irrigation or in poorly drained soils, some or all of the soil profile may be completely saturated with water. Under these conditions, since there is no net change in water content, the flux, F_s, of water is dominated by the difference in gravitational force driving water through the soil $(\Delta\psi_g)$ and the hydraulic resistance (Z_r) of the soil to the flow of water, that is

$$F_s = \frac{\Delta\Psi_g}{Z_r} \qquad (3.9)$$

The hydraulic resistance of a uniformly saturated soil depends on the size and configuration of the soil pores which, in turn, are a consequence of soil texture and structure.

From the above we can see that, because water in a fully saturated soil is held with a minimal tension, the driving force for the vertical movement of water, i.e. the hydraulic gradient, is simply the difference in height of water above and below the soil column. The volume of water moving down the column depends on this force as well as on the hydraulic resistance of the soil. Typical flow rates in saturated soil are indicated in Table 3.1.

HYDRAULIC RESISTANCE. This depends on the configuration and size of the particles that make up the soil because the total flow rate through soil pores is proportional to the fourth power of pore radius. Flow is therefore much greater: (i) in soils with large pores, and (ii) where the structure is made up of uncompacted and granulated particles. So, the value of Z_r is less in uncompacted sandy soils than in compacted and/or clay soils.

UNSATURATED FLOW. In saturated soils, the rapid movement of water is through the largest, most continuous pores. In unsaturated soils, such *macropores* are filled with air and water has to move sluggishly through finer pores and along a more tortuous path. As a result, water movement becomes progressively slower in unsaturated soils as hydraulic resistance increases in response to the lowering of ψ_m. Unsaturated flow is

therefore governed by the same factors as in saturated flow, i.e. the driving force and the hydraulic resistance. However, in unsaturated soils the dominant factor is the matric potential gradient, i.e. the difference, $\Delta\psi_m$, between the matric potential of the moist soil areas and the drier areas to which the water is moving.

WATER VAPOUR MOVEMENT IN SOILS. Water vapour can either move internally through the pore spaces within a soil or can be lost to the external atmosphere via evaporation from the soil surface. Water vapour moves within the soil in response to differences in vapour pressure. Consequently, water vapour will move from an area of moist soil with a high vapour pressure to a drier region of soil where the vapour pressure is lower. Similarly, water vapour in a uniformly moist soil will move from a region of higher temperature (high vapour pressure) to one of lower temperature (low vapour pressure).

3.2.3 Characteristics of water movement in soil

The above discussion has described the major forces that influence the state and movement of water in soils and their combined effect on the soil water balance. We now consider how these interact with plant systems to determine the fraction of the soil water balance at any particular site that can be accessed for crop production.

RUNOFF, EVAPORATION, DEEP DRAINAGE. If we imagine a cropped soil which has been uniformly wetted by rainfall or irrigation and then allowed to dry, there are a number of stages that can be identified in terms of the

Table 3.1. Typical flow rates in different saturated soils.

Flow rate (cm h^{-1})	Soil type
Very slow (< 0.25)	Shallow soils low in clay and/or organic matter
Slow (0.25 – 2.5)	Shallow soils with a high clay content but usually still low in organic matter
Medium (1.25 – 2.5)	Silts and loams
Fast (> 2.5)	Deep sands and aggregated silt loams

energy–soil moisture relations of the soil as it progressively becomes depleted in moisture. The most important of these characteristics with progressive drying of the soil are described below.

WATERHOLDING CAPACITY. When a soil has been fully wetted and drainage begins, it is saturated with respect to water, i.e. it is at a water content equal to its saturation ratio (ϕ_s). At this stage, the volume percentage of water in the soil is known as the *waterholding capacity* of the soil.

FIELD CAPACITY. In a saturated soil, the rapid drainage of water in response to the hydraulic gradient ceases after two or three days and now the soil is said to be at *field capacity*. This occurs when water in the macropores has been replaced by air. At this stage, the matric potential of the soil will be in the range −0.01 to −0.03 MPa. This range reflects the transitory nature of field capacity. Its further movement is slow because this unsaturated flow is now due primarily to capillary forces. For a given amount of rainfall without any runoff, field capacity controls the depth of water penetration.

PERMANENT WILTING PERCENTAGE. Evaporation from cropped soil occurs through the twin processes of transpiration and direct evaporation from the soil surface. At some stage, if water in the soil profile is not replenished by rainfall or irrigation, plants begin to wilt in response to water stress. Plants which remain in this condition are said to be at *permanent wilting point*. Although this is rather a crude concept, for many crops, the permanent wilting point does often correspond to a soil water potential of between about −1.0 and −2.0 MPa, with a mean of 0.15 MPa (Jenny, 1980). The soil water content at this stage is called the *permanent wilting percentage*. Water remaining in the soil is now too tightly held within *micropores* and around soil particles to be available to higher plants. We return to the subject of plant water status later in this chapter.

HYGROSCOPIC COEFFICIENT. As soil moisture falls below the permanent wilting point, the remaining water molecules are adsorbed to the surfaces of soil colloids. This water is held so tightly that it can be considered as a non-liquid and can only move in the vapour phase. The moisture content in the soil at

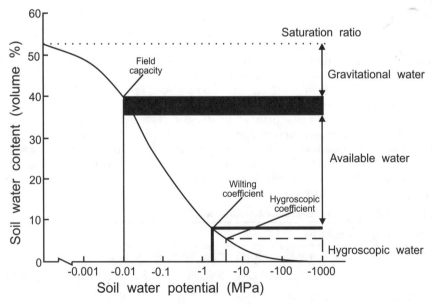

Fig. 3.3. Different terms to describe the water status of soil as a function of soil water content and potential. (Adapted from Brady and Weil, 1999.)

this stage is called the *hygroscopic coefficient* and is correspondingly greater in clay soils that are rich in colloidal matter than in sandy soils which are low in colloids.

AVAILABLE WATER. Figure 3.3 shows the generalized shape of the relation between soil water content and soil water potential with representative ranges for field capacity, wilting point and the hygroscopic coefficient. In principle, plant available water is simply the difference between the amount of water in the soil at field capacity and that at the permanent wilting point. However, the presence of high levels of salts in the soil solution can influence the availability of water to plants. These osmotic effects reduce the amount of water available to plants by increasing the amount of water left in the soil at the permanent wilting point. In most humid soils, osmotic potential effects are negligible but they can have a profound influence in some soils of arid and semi-arid regions, especially those with high salt contents. However, in most circumstances the available water capacity (AWC) of a soil can be taken as the difference between field capacity and permanent wilting point.

SOIL DEPTH. In many circumstances, it is not the soil texture or structure that determines the amount of water that is available to plants. Rather, it is the ability of plant roots to descend to deeper levels in the soil profile and the depth of the soil profile itself. Soil depth is an important factor in many tropical soils, particularly shallow soils or those which have impenetrable layers of material within the soil profile. To provide a basis for comparison between different soil types it is usual to present AWC in terms of millimetres of water per metre depth of soil. Of course, where the soil depth is known to be less than 1 m the AWC should be adjusted accordingly.

SOIL AIR. Soil aeration controls soil levels of O_2 and CO_2. These two gases influence the respiration of roots and soil microorganisms. For respiration to continue in the soil, O_2 must be supplied and CO_2 removed. For most land plants the supply of O_2 in the soil

air must exceed 10% (Brady and Weil, 1999). Poor soil aeration in field soils occurs under two conditions: (i) when moisture content is too high to allow any room for gases; (ii) when the exchange of gases with the atmosphere is too slow, even if sufficient air space is available in the soil.

EXCESS MOISTURE. The extreme case of excessive soil moisture is *waterlogging*. This condition occurs at low spots in the field or when drainage is impaired by poor soil structure. Even well drained soils can become waterlogged during a heavy rainstorm or when excessive irrigation has been applied. The compaction of soil caused by ploughing or the use of machinery can also encourage waterlogging because the reduction in porosity caused by compaction in turn reduces the amount of water that can be absorbed before saturation.

GASEOUS EXCHANGE. The faster the removal of O_2 and the corresponding release of CO_2, the greater is the demand for exchange of gases between the soil and the atmosphere. This exchange occurs through two mechanisms: mass flow and diffusion. Mass flow is the result of fluctuations in soil water content. For example, as water moves into a soil after rainfall or irrigation, soil air is forced out. Similarly when evaporation occurs from the soil or when water is extracted by roots, air is drawn into the spaces that are left behind. Most gaseous exchange in soils occurs through the process of diffusion. Here, each gas moves in a direction determined by its own partial pressure, i.e. along a partial pressure gradient. (The partial pressure of a gas in a mixture is simply the pressure that it would exert if it were present alone in the volume occupied by the mixture.) Diffusion allows movement of gases from one zone of soil to another even though there is no overall pressure difference for the total mixture of gases. Because the movement of each gas is in response to its own partial pressure gradient there may be differences in the concentrations of gases in different soil layers. For example, although soil air contains about the same level of nitrogen as the air above it (78%), it is consistently lower in O_2

and higher in CO_2. The O_2 content in the upper soil layers may be only slightly below that in the ambient atmosphere (20%) but may decline to below 5% in lower horizons of poorly drained soil with few macropores (Brady and Weil, 1999). Soil levels of CO_2 vary inversely in proportion to the amount of O_2 present and at levels above about 10% can be toxic to some plant processes. From the above it can be seen that, as a rule of thumb, O_2 levels must be above 10% and CO_2 levels below 10% in a soil for plant function not to be impaired. For many tropical soils, there are the twin dangers of high CO_2 levels, associated with increased respiration rates in the warm soil environment, and low O_2 levels, caused by transient waterlogging due to the intensity of any rain that does fall.

3.3 Plant water energy

In a metabolically active plant, cells are composed of more than 80–90% water. This water forms a liquid continuum from the surface of roots through the plant's tissues to the leaf surfaces from where it evaporates. The passage of water through the soil, plant and atmosphere is often considered in terms of a soil–plant–atmosphere continuum (SPAC). Between its uptake from the soil and its loss to the atmosphere, plant water performs a number of functions, including the transport of nutrients and metabolites, and contributes to structural integrity through the turgidity of various organs. Unlike soil water, the water content of plant tissues must not only remain high for the maintenance of metabolic functions but can only fluctuate across a fairly narrow range; a change of water content of 20–25% of the value at maximum hydration results in the cessation of most growth processes (Slatyer, 1967). A further complication is that, whereas the biological activity of a dry soil can be resumed on its rehydration, any depletion in the water content of plant tissues below a critical threshold results in their death. Further growth of the plant must then depend on those tissues that have survived dehydration or on the production of new

plant material. From the above, it can be seen that the ability of a plant to buffer its water status against the vagaries of soil water supply and atmospheric demand is critical to its success and survival.

Later in this chapter, we discuss the efficiency with which plants capture water through their roots and transpire much of this through their leaves in exchange for CO_2. Before then, we need to consider the regulation of water content and status within plant tissues.

3.3.1 Plant water potential and content

As with soil, plant water can be considered in terms of energy and content. Water movement within plants occurs along gradients of free energy which depend on the difference in plant water potential, ψ_p, from regions of high energy (abundant water) to those where the free energy is lower (less water). The water potential within plant tissues is always less than that of pure water because of the influence of solutes on the osmotic potential (ψ'_o) and matric forces (ψ'_m) caused by the bonding of hydrogen ions to cell walls and proteins. Differences in height within the plant should mean that gravitational forces also influence water potential. However, except in very tall trees this effect is negligible. Unlike soil elements, the outward force exerted by turgid plant cells provides an additional, positive, hydrostatic or turgor pressure (ψ_h). For plant systems the resulting water potential can be expressed as

$$\psi_p = \psi_h + \psi'_o + \psi'_m \qquad (3.10)$$

From the earlier discussion of soil water energy, we can see that the water potentials of soils and plants are always less than those of pure water, and for plants to extract water from the soil, ψ_p must always be less (i.e. more negative) than ψ_s.

In practice, plants extract water from the soil in proportion to the distribution of their active root systems. This means that some parts of the soil may remain close to field capacity (between about 0.01 and 0.03

MPa) whilst other regions approach the permanent wilting percentage (between about −1.0 and −2.0 MPa). During daylight, water movement within the plant depends on the relative values of ψ_p across different parts of the plant, which in turn depend on the loss of water through stomata on the surfaces of leaves. At night, when stomata are closed, ψ_p returns to equilibrium with the soil (i.e. $\psi_p = \psi_s$).

RELATIVE WATER CONTENT. Often, a decrease in the actual water content of tissues has a greater effect on plant metabolism and growth than any changes in ψ_p or ψ_h (Sinclair and Ludlow, 1985). Tissue water content can be expressed as the amount of water per unit dry weight, per unit fresh weight and per unit weight of water at full hydration (i.e. the leaf relative water content, RWC). For our purposes, RWC is a most useful means of quantifying plant water deficits, since it is not influenced by changes in tissue dry weight (Slatyer, 1967; Turner, 1981). The value of RWC can be estimated by measuring the fresh weight (W_2) of leaf samples (usually cut as small discs) and rehydrating these on a water surface to obtain their saturated weight (W_3). The discs are then oven dried to obtain their dry weights (W_1). RWC can then be calculated from:

$$\text{RWC} = \frac{W_2 - W_1}{W_3 - W_1} \times 100 \qquad (3.11)$$

3.3.2 Water movement in plants

The movement of water through the plant components of the SPAC from soil to atmosphere is often considered as a continuous transpirational 'stream'. However, even when soil water supply to roots is unrestricted, diurnal changes occur in the water potential of plant shoots because leaf surfaces must lose water through transpiration. The decrease in ψ_p of the shoot is caused by the hydraulic resistance, Z_p, to water movement in the soil–plant transpirational stream. We can simplify the flux, F_p, of water through the plant system as

$$F_p = \frac{\Delta \Psi_p}{Z_p} \qquad (3.12)$$

where $\Delta\psi_p$ is the gradient of water potential across the plant. Often, a linear relationship (equivalent to Z_p) exists between F and $\Delta\psi_p$, especially across moderate-to-high fluxes (Koide *et al.*, 1994) because osmotic forces usually play only a small role in water movement. Here, we can assume a steady state in which water absorbed by the root system is equal to that subsequently transpired by the canopy. However, such conditions may not prevail where much of the plant material consists of young, actively growing tissues that require water for cell expansion. Similarly, many plant organs, such as leaves, stems and fruits, may act as 'capacitors' which change in size as a consequence of absorbing or losing water. Clearly, dehydration and loss of cellular water in response to soil water deficits may result in a temporary imbalance between the loss and gain of water by the whole plant.

3.3.3 The soil–plant–atmosphere continuum (SPAC)

From the above it is clear that the movement of water through soils and plants is part of a continuum between the soil and the atmosphere since all plants must lose water through their leaves and replenish this by extracting water from the soil through their roots. Figure 3.4 summarizes the various physical and physiological factors that act on the SPAC.

3.4 Evaporation

All atmospheric moisture originates from the earth's surface, where water in liquid or solid phase is transformed into water vapour that can be transported in upward air movements. The water vapour originates from the process of *evaporation* which itself has two components. The first is the evaporation that occurs *directly* from water surfaces or from land areas with soil or vegetation cover that has recently been wetted by precipita-

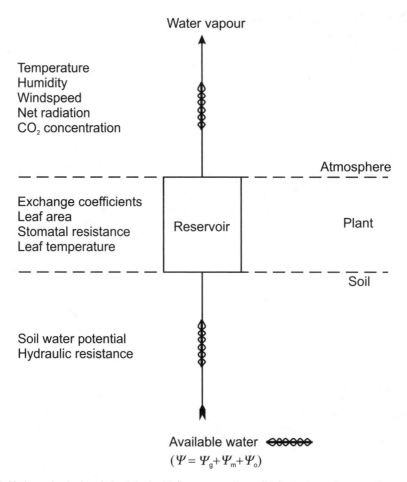

Fig. 3.4. Various physical and physiological influences on the soil–plant–atmosphere continuum.

tion. The second source is *transpiration* which is physically the same process as direct evaporation but occurs through the stomatal apertures of plants which draw up water through their root systems.

Direct evaporation from the soil surface (E_s) and transpiration (E_p) are collectively known as *evapotranspiration*, E_t, which is a measure of the total flow of water vapour into the atmosphere. There are three physical factors that influence E_t. The first is the capacity of air to hold water vapour, i.e. the atmospheric 'demand' for water. This quantity is a function of the temperature of the air and, as a consequence, warm, tropical air masses can absorb more moisture than cold, temperate air. The actual amount that can be

absorbed by the atmosphere depends on relative humidity at any particular temperature, i.e. the amount of water that is already there. The lower the relative humidity the greater is the potential for further evapotranspiration. The second factor that affects E_t is the amount of energy available for the latent heat used in evaporation and transpiration, i.e. the energy required to convert liquid water into a vapour at the same temperature. This energy is mainly provided by solar radiation. Thirdly, air turbulence is required to replace saturated air above the crop with unsaturated air from surrounding areas.

In addition to the physical factors mentioned above, plant communities exert a physiological influence on the transpirational fraction of E_t through a combination of stomatal regulation and the total evaporative surface area of their leaf canopies. When a crop fully covers the ground and its roots are well supplied with water, there is minimal stomatal control over transpiration, and evaporation directly from the soil surface is negligible. In these circumstances, evaporation proceeds at a rate determined by the atmospheric demand for water vapour. The evaporation rate is therefore the maximum for these conditions and this is known as the *potential* evaporation E_0. The value of E_0 varies by 10–20% among different crops owing to differences in colour, height, aerodynamic characteristics and stomatal control (Loomis and Connor, 1992). Often, short green grass fully covering the ground and with an unrestricted supply of water is used to calculate the *reference evapotranspiration*, E_{to}, which is used as a standard measurement of environmental demand at any location (Doorenbos and Pruitt, 1977). For individual crops, E_0 can be related to E_{to} by a *crop coefficient*, K_c. Representative K_c values for a series of different crops and locations are presented by Doorenbos and Kassam (1986). Although these are empirical functions which have been calibrated for local conditions, the general principle is that K_c values for any particular crop increase with the degree of crop cover through the season as the ratio $E_p : E_s$ increases until E_p approximates to E_t. For most short arable crops with full cover and sufficient water, $E_0 = E_{to}$, i.e. $K_c = 1$.

From the above we can see that where the supply of water is unrestricted, e.g. over open water surfaces and moist vegetation, evaporation occurs at the rate set by the potential demand of the atmosphere. Thus, in these circumstances, E_t can be calculated from the energy balance equation

$$R_n = \lambda E_t + C + H \qquad (3.13)$$

where, R_n is the net radiation, i.e. the difference between the incoming and outgoing radiation. The term λ, represents the latent heat of evaporation, i.e. the amount of energy required to convert water from a liquid to a gas without a change in temperature; H is the amount of heat transferred to the soil and C is the loss of heat by convection. Measurements of R_n are rarely available from meteorological stations but values can be derived from solar radiation, or sunshine hours, temperature and humidity.

In practice, for most tropical crops evaporation is often limited by the supply of water to the roots, and in arid climates E_{to} often substantially exceeds E_t because stomatal closure for part or most of the day restricts transpiration. In these circumstances, the energy balance approach and the use of empirical K_c functions fail for much, if not all, of the growing season. To devise rational agronomic strategies for these water limited environments, the tropical agronomist must look more closely at the physical and physiological mechanisms that determine plant (E_p) and soil (E_s) evaporation and must understand the link that crop productivity has with evaporation, or more precisely with transpiration.

3.4.1 Transpiration

To photosynthesize, plants must absorb CO_2 from the atmosphere through stomatal apertures on the surfaces of their leaves. An inevitable consequence of this CO_2 uptake is that water is lost by transpiration through the same pores. Furthermore, only about 1% of the energy absorbed by leaves is used for photosynthesis; the rest tends to heat the leaves. To prevent the temperature of their leaves increasing throughout the day, plants must dissipate the excess energy either by transpiration, convection or radiation. (The flux of water vapour from the leaf to the atmosphere, corresponding to the photosynthetic uptake of CO_2 is illustrated schematically in Fig. 2.3.)

The link between the amount of CO_2 entering a crop and the amount of H_2O leaving it means that, at least in principle, seasonal transpiration can be used to estimate the carbon assimilation of a crop. This approach has advantages in rainfed tropical environments where it is the shortage of water rather than the amount of solar radiation that determines crop productivity. Also, in many circumstances, stomatal closure means that little CO_2 enters or H_2O exits from leaves even when they are intercepting solar radiation. Therefore, when water stress restricts crop growth, a knowledge of the potential conversion efficiency of radiation may be of little predictive value and it may be more appropriate to consider the relation between crop productivity and transpiration. However, to establish this link we first need to distinguish between the contribution that E_p and E_s make to E_t from a cropped field.

3.4.2 Evaporation from the soil surface

When crops cover the ground for much of the season, E_s may be a small fraction of total evaporation and can effectively be ignored in calculations of crop water use. However, this is rarely the case in tropical environments where: (i) there is usually a large proportion of bare ground because crops never achieve full cover and/or (ii) the soil surface is frequently rewetted by relatively small amounts of rainfall. On each occasion, this rainfall may be insufficient to penetrate more than the uppermost soil layer and therefore enter the root zone. In these circumstances, E_s may dominate estimates of total evaporation. For example, Cooper et al. (1983, 1987a, b) showed that for unfertilized

barley crops growing in a Mediterranean environment, E_s was as much as 75% of total evaporation.

Evaporation from bare soil proceeds in two distinct stages (see Fig. 3.5). In stage 1 (energy dependent), the evaporation rate is primarily determined by the net radiation (R_n) at the soil surface and, during this period, E_s is similar in magnitude to E_o. However, an unrestricted water supply to the soil surface rarely continues for more than a few hours, and a rapid decline in E_s occurs (stage 2, supply dependent) once the soil surface layers begin to dry. Following the rapid transition to stage 2, E_s progressively declines in inverse proportion to the square root of time. This is because water vapour must be drawn from increasing depths in the soil profile and must overcome the large hydraulic resistance (Z_s) of unsaturated soil.

The above characteristics describe the behaviour that is typical of bare soils after rainfall or irrigation. However, the crop canopy, or more correctly the ground covered by a crop canopy, will influence both the duration and the magnitude of E_s. This is because the crop shades a proportion of the ground surface and roots extract moisture from the soil surface layers. Thus, the crop influences both the energy supply at the soil surface and the soil water supply required for E_s to proceed at the theoretical rate for bare soil.

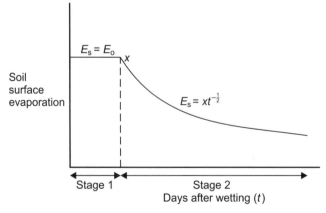

Fig. 3.5. Evaporation from a wet soil showing stage 1 drying (soil evaporation, E_s, is equal to the potential evaporation, E_o) and stage 2 drying (E_s decreases in inverse proportion to the square root of time, t).

Water lost from the soil surface is of little practical benefit to growing crops. Therefore, where E_s is a significant component of total evaporation it must be taken into account when calculating crop yield responses to rainfall or irrigation. Physical measurements of E_s are laborious and indirect methods often rely on estimates of the proportion of ground covered by the crop – which is itself a difficult factor to quantify. However, the omission of E_s from calculations may lead to large variations in published estimates of crop water use.

3.5 The fate of water in crop systems

The preceding discussion considered what proportion of the rain that falls at any location ends up being used by plants to produce dry matter and ultimately yield. As with radiation (see Chapter 2), the overall efficiency, ε, can be described by

$$\varepsilon = \varepsilon_u \, \varepsilon_o \, \varepsilon_l \, \varepsilon_c \, \varepsilon_e \, \varepsilon_w \, \varepsilon_q \, \varepsilon_r \, \varepsilon_H \qquad (3.14)$$

where ε_u, ε_o and ε_l refer to the proportions of water left in the root zone after deep drainage, runoff and lateral movement respectively, ε_c is the capture efficiency of water by plant roots, ε_e is the fraction of total evaporation not lost directly from the soil surface (i.e. transpiration), ε_w is the efficiency of dry matter production per unit of transpired water, ε_q is the photochemical efficiency i.e. the energy value of carbohydrates, proteins and lipids, ε_r is the respiration efficiency and ε_H is the yield efficiency.

3.5.1 Availability of resource (ε_u, ε_o and ε_l)

From Fig. 3.1, it can be seen that the amount of water in the soil depends on seasonal precipitation minus losses through runoff, drainage and lateral movement. Unlike radiation, where energy previously incident cannot be stored for future use, plants may also have access to water stored in the soil, even before the crop was sown. Another important point is that the amount of water in the soil is not synonymous with that available to plant roots because of the energy considerations mentioned earlier. Further, the amount of soil water that can be extracted at any particular time by a crop depends on the characteristics and behaviour of its root system.

3.5.2 Capture of resource (ε_c)

The amount of available soil water that can be extracted by a crop depends on the characteristics and behaviour of its roots. However, root growth is complex and poorly understood; *in situ* measurements of root systems are difficult and time consuming. Consequently, a knowledge of the underlying plant and soil factors that govern root growth is still mostly missing. However, for our purposes three features of root systems are critical in determining their capture of water.

First, there is the speed at which the root system extends into soil. This can be considered in terms of a *root front velocity* which depends both on the species in question and the physical characteristics of the soil. Second, there is the maximum depth to which the root system extends, which again depends on species and soil. The depth of roots allows us to calculate the maximum amount of water that is available to the crop throughout its growing season. The third feature is the length of root per unit volume of soil (l_v). This determines the rate and amount of water at any given depth that can be taken up by plant roots. Although, there may be differences in the ability of different segments of the root system to take up water, it is generally assumed that inflow rates are a uniform function of l_v. Gerwitz and Page (1974) reviewed the effects of soil moisture, fertility, plant age and mass on the growth of plant root systems. They concluded that a negative exponential function usually gave a satisfactory description for most cases. So, if W_r (kg dry matter of roots per m^2 of ground) is the root mass and z (m) is the distance below the soil surface at $z = 0$, then the function

$$\beta W_r\, e^{-\beta z} \qquad (3.15)$$

gives the root density per unit volume of soil at depth z, where W_r is in units of kg m^{-3} and β (m^{-1}) is a constant. Integrating Equation 3.15 with respect to z between $z = 0$ and z gives

$$W_r\, (1 - e^{-\beta z}) \qquad (3.16)$$

Thus, $1 - e^{-\beta z}$ gives the fraction of root mass that lies between the soil surface and depth z.

3.5.3 Conversion of resource (ε_w)

For any crop, the relation between total dry weight and seasonal transpiration is often linear with a slope, ε_w (g kg^{-1}), known as the *dry matter/transpired water ratio* (Squire, 1990). Steduto (1996) provides a comprehensive review of the link between dry weight and transpiration. The closeness of this link is probably because both dry matter accumulation and transpiration are closely related to the solar energy received per unit field area and the fraction of incident energy intercepted by a crop (Monteith, 1981, 1986a). The value of ε_w does not seem to be seriously affected by nutrients (de Wit, 1958; Wong *et al.*, 1979; Tanner and Sinclair, 1983) or water stress (Azam-Ali *et al.*, 1994). However, when the same crop is grown at different sites or seasons, the value of ε_w often varies between sites and/or seasons. There are three principal causes of this apparent variation in ε_w.

The contribution of roots to total productivity

When crops are well supplied with water, the contribution of roots to the total dry weight is usually small and can often be ignored in calculations of ε_w. However, where crops are grown on limited soil moisture, root weight may be a substantial component of total biomass and root : total weight ratios of between 0.3 and 0.5 have been reported for a number of crops including sorghum (Myers, 1980), barley (Gregory *et al.*, 1984), millet (Azam-Ali *et al.*, 1984)

and groundnut (Nageswara Rao *et al.*, 1989a). A further complication is that root : total weight ratios are strongly time dependent with proportionately less assimilate allocated to roots late in the season. Nevertheless, despite their importance, because measurements of root weight are extremely time-consuming, most calculations of ε_w are based on the above-ground component of crop weight. Clearly, the inconsistency of using total (i.e. including roots) or only above-ground dry weight in calculations of ε_w may lead to marked variations in published values both between and within experiments where the fractional weight of roots changes during the season and where experimental treatments include differences in soil moisture (see Simmonds and Azam-Ali, 1989). For example, for three populations of groundnut growing on stored soil water in India, Azam-Ali *et al.* (1989) showed that the calculated value of ε_w increased from 1.57 g kg^{-1} to 3.00 g kg^{-1} when roots were included in measurements of crop dry weight.

Even when root weights are included in calculations of ε_w, measured values are prone to substantial errors. This is often because of difficulties in extracting roots from soil, selecting representative samples of the soil volume and, often, the poor replication of sampling in space and time compared with measurements of above-ground components of crop weight.

Of course, any underestimates in the calculation of total crop dry matter because root weight has been excluded will also lead to variations in the calculated values of ε_s (see Chapter 2) as well as ε_w. However, it is appropriate to mention the contribution of roots in this section as their relevance becomes greater when the supply of soil moisture limits crop productivity.

The influence of atmospheric saturation deficit

Even when crops are suffering from severe water stress, the air in the sub-stomatal cavities of their leaves is saturated at the leaf temperature. The driving force for evaporation is therefore always the difference between the saturation vapour concentration in the leaf and that in the air. As a consequence, the value of ε_w is inversely related to the saturation deficit (D) of the atmosphere during the season (Tanner and Sinclair, 1983). It is not the absolute atmospheric moisture content but the difference between the saturation vapour pressure and actual vapour pressure of the air, that is important. When leaf and air temperatures are the same, the driving force for evaporation is the true saturation deficit of the air, i.e. D, and in these circumstances D sets the potential rate for transpiration. Strictly, the driving gradient for transpiration depends on leaf temperature as this sets the saturation vapour concentration at the leaf surface. However, measurements of leaf temperature are rare and in most circumstances D is an acceptable approximation. Similarly, although transpiration at any time is closely linked to the current saturation deficit, the seasonal mean value of D is often an acceptable approximation because changes in D during the season are usually small compared with differences between seasons (Monteith, 1986b).

The theoretical influences of root weight, surface evaporation and saturation deficit on calculations of ε_w are shown in Fig. 3.6. When root weight and E_s are included, they improve the general applicability of ε_w. Further, when D is included as a normalizing factor (Monteith, 1990), the value $\varepsilon_w D$ (g kPa kg^{-1}) appears to remain conservative throughout the season (Squire, 1990) and, unlike ε_w, remains fairly constant from site to site and year to year. Therefore, $\varepsilon_w D$ may be a more useful index of crop productivity than either ε_s or ε_w, especially when the supply and/or demand for water is limiting.

For any particular crop, the value $\varepsilon_w D$ can be treated as a quasi-constant in terms of yield prediction. Strictly, the correct terminology for $\varepsilon_w D$ is 'the product of the dry matter/transpiration ratio and the mean seasonal saturation deficit'. However, a simpler term is the 'transpiration equivalent', Ω_w (see Azam-Ali *et al.*, 1994).

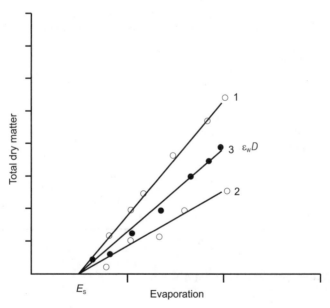

Fig. 3.6. Schematic examples of how total crop dry matter including roots is linearly related to transpiration: (1) when the atmosphere is wet; (2) when the atmosphere is dry; (3) when the dry matter : transpired water ratio, ε_w, is multiplied by the mean saturation deficit throughout the season (D) to provide a constant value for the transpiration equivalent ($\Omega_w = \varepsilon_w D$). E_s indicates that the soil surface evaporation must be deducted from total evaporation to provide linear fits between dry matter and transpiration.

3.6 Conclusions

In this chapter, we have seen that it is the gradients in soil and plant water potentials, rather than water content *per se*, that determine the flux of water through soils and plants. The loss of water to the atmosphere occurs through the process of transpiration whose flux is determined by a combination of physiological and physical factors acting on the soil–plant–atmosphere continuum. A knowledge of seasonal transpiration and the transpiration equivalent (Ω_w) of a particular species is a good indicator of its total productivity. In fact, in circumstances where water is the most limiting resource, estimates of productivity based on a knowledge of transpiration may be more reliable than those based exclusively on a knowledge of accumulated intercepted radiation. However, transpiration-based estimates of productivity need to account for the contribution of roots to total dry matter, the component of evaporation lost directly from the soil surface (i.e. E_s) and the influence of saturation deficit (D) to allow for variations in the value of ε_w between sites and seasons.

Despite their potential use in predicting crop productivity, estimates of transpiration are far more difficult to obtain than comparable estimates of the combined evaporation, E_t, from crop (E_p) and soil (E_s) surfaces. Nevertheless, if E_s can be measured or estimated, then a knowledge of the total evaporation at any site can be used to predict the total dry matter, W, of a crop, i.e.

$$W = \frac{(\Omega_w)(\lambda E - E_s)}{D} \qquad (3.17)$$

and from Equation 2.9,

$$W = \varepsilon_s \int fS_0 dt = \frac{(\Omega_w)(\lambda E - E_s)}{D} \qquad (3.18)$$

Where the reproductive fraction of total dry matter can be reasonably predicted from a knowledge of the total productivity, grain yield, G, can either be calculated as a function of solar radiation when water is not limiting, hence

$$G = H(\varepsilon_s \int fS_0 dt) \qquad (3.19)$$

or transpiration when water is limiting,

$$G = H\left(\frac{(\Omega_w)(\lambda E - E_s)}{D}\right) \qquad (3.20)$$

There are very few estimates or reports of Ω_w for any of the major crops of the world and it is therefore difficult to compare the relative merits of water-based and radiation-based estimates of crop productivity. Clearly, there are special circumstances in which either the supply of water or the interception of solar radiation alone dominate the process of dry matter production by crops. However, in most cases crops are continuously responding to short-term changes in both solar radiation and water supply with neither factor being optimal. Consequently, to successfully predict the performance of crops throughout their growth and across the widest possible range of circumstances, agronomists should attempt to take account of both light- and water-driven productivity.

In these brief discussions of light-driven and water-driven methods of calculating crop productivity, it is important to note that it is the method by which they predict total biomass that is relevant. Irrespective of how well the water-driven method may or may not estimate productivity, it still suffers from the same limitations as the solar-driven method in terms of the subsequent influence of factors such as respiration, photochemical composition and the fraction allocated to harvestable yield where this is the most important component of crop biomass.

4

Genotype (What?)

Genetics and breeding have provided agronomy with a range of crop types. Continued modification of traits such as adaptation to season, pest resistance and product quality have allowed mankind to maintain food production at levels that equal or surpass the increasing demand by the growing world population. Genetic improvement has also led to very great changes in the way land is used. Early flowering cereal cultivars have encouraged arable practices to move into previously pastoral land, while changes in the chemical quality in the products of some tree crops has given rise to the vast perennial plantations of South-east Asia. Such improvements in plant performance need not always lead to an extension of agricultural area, however. Any increase in maximum yield per unit area provides an opportunity to reduce the area of land under any crop, potentially allowing a diversity of crop types, crop systems or semi-natural vegetation to occupy the land.

Genetic improvements alone have rarely altered yield. Rather, they have created a potential which agronomy has acted on (Simmonds, 1979). As Evans (1993) has demonstrated, the systematic and widespread increases in agricultural area and yield throughout history have resulted from the combined effects of genetic modification and agronomy. Even when these disciplines have not been coordinated, one has made the way and the other followed. Such is their interdependence that in most instances of agricultural improvement, their individual contributions to any particular historical increase in the yield of a crop are indistinct.

Some of the most successful crop production enterprises have combined breeding and agronomy from their inception, as the development of oil palm research in Malaysia in the 1960s and 1970s demonstrated. Too often though, genetics and breeding have sat uneasily with physiological agronomy. Agronomists have been presented with a group of new cultivars that have to be fitted into local seasons and practices. The cultivars might have been bred firstly for resistance to pests, so their response to soil and weather is unknown or considered secondary. Breeders have often argued that physiological knowledge only explains with hindsight how breeding has increased yield; it has rarely directed the search for traits. Despite some truth in such statements, there would seem to be no case now or in the future for the exclusion of one discipline by the other. New genetic types are now being distributed around the world with such speed that repetitive trialling at each potential site is simply not feasible. For efficiency, the crop types have to be characterized in advance in terms of some traits describing the way they interact with the physical environment; and once in the ground, their performance has to be defined and compared by simple and universal trait variables.

The main part of this chapter attempts to examine broadly how the crop types and varieties resulting from natural selection and breeding differ in the way that they intercept and use environmental resources. It is to the advantage of agronomy that these differences are concentrated among certain physiological traits, while other traits are much less affected by selection. Knowledge of where differences (and the absence of differences) are likely to be should increase the rate at which new cultivars are characterized and introduced. First, some background is given on the origins and genetic improvement of crops.

4.1 Evolution and genetic manipulation

Fewer than 20 plant and eight animal species account for more than 90% of the world's food. The main food crops were originally developed by Neolithic Stone Age farmers and few new or previously underused species have achieved global prominence in modern times. The main staple crops (Table 4.1) such as wheat, maize, rice and potato all have characteristics that have

made them successfully able to respond to market forces. They are each highly variable genetically and can be adapted to grow in a range of environments. They have all been amenable to selection and breeding. Their products are relatively easy to store and prepare. Moreover, they all have a high potential yield and respond to efficient management. Despite relying on a narrow cohort of the main crop types, agriculture also uses many other locally important species that yield grain, fruit, beverages, pharmaceuticals and industrial materials. The minor food species are found more commonly in severe dry or cold environments where the main staple species offer no advantage, or else in gardens where the aim is to augment a carbohydrate diet with protein crops and fruit and vegetables. In the tropics, subsistence gardens and peri-urban systems of cultivation commonly display a wide range of species. Coconut is ubiquitous, all its parts being used for either building material, food, feed, drink or remedies. A few minor crops have even become economically important on a national or regional scale. The use of beverage crops, such as tea and coffee, goes back several thousand years at least but their rise as plantations covering substantial productive areas of Asia and Africa is a phenomenon of the last 100 years or so. The even more recent expansion of rubber, cocoa and oil palm was stimulated variously by demands from industry or consumers outwith the tropics.

Thousands of other plants have been used as crops or as wild sources of food and material, but mostly on a local scale or at a low intensity of production. These rare crops have been well documented in some parts of the world. For example, Burkhill's (1935) dictionary of the economic products in the Malay Peninsula has entries in its >2000 pages of detailed text for maize, rice, oil palm and rubber but also for numerous tree and herb species that are all but obscure, except to those who live off the forest itself.

Tropical agriculture therefore includes a great diversity of plant form and product. There are arguments, however, that mankind relies on too narrow a range of species and that this is leading to reduced

Table 4.1. Area and total production of the major crops of the world. (From FAO, 1998.)

Species	Area under cultivation (1000 ha)	Global production (1000 Mt)
Wheat	224,374	588,842
Rice	150,305	563,188
Maize	137,430	604,013
Soybean	70,690	158,327
Barley	61,699	138,820
Coconut	47,696	241,961
Sorghum	44,442	63,451
Millet	37,596	29,204
Haricot bean	25,689	17,619
Rapeseed	24,987	33,568
Groundnut	23,799	30,972
Sunflower	21,251	24,942
Sugarcane	19,438	1,252,266
Potato	17,949	295,632
Cassava	16,188	158,620
Oat	14,580	25,802
Chickpea	11,194	8,587
Rye	10,810	20,977
Sweet potato	8,867	129,164
Sugarbeet	6,955	258,859

biological diversity and the extinction of some crop species and genotypes. Bambara groundnut (*Vigna subterranea*), for instance, is an ancient and valuable leguminous protein crop, whose growing area and availability, specially in West Africa, has been reduced during the present century, partly as a result of the expansion of introduced crops such as groundnut and haricot beans (Doku, 1997). Bambara groundnut is drought tolerant, nutritious and tasty, so the reasons for its decline are not straightforward (see final chapter). Evans (1993) redresses the balance in favour of the main crops by indicating the greatly increased range of genotypes that are extant within each of these main crop species, and calls for greater concentration of research into their biology.

We return to the discussion of main crops versus minor crops in the final chapter, but draw attention here to the following. If conservation and continued or increased use of minor crops is the aim, then their agronomy has to be improved as well as their genetic material conserved. In the case of bambara groundnut, people have lost not just the continuity of seed but also the local knowledge of how to grow it. This is not just a tropical problem. Much the same can be said of a range of minor cereal and industrial crops in Europe.

4.1.1 Origins, dispersal and conservation

Several authors, following Vavilov (1928), have argued that many of the staple crops can be traced to a geographical location at which they originated or diversified. However, the origin and early evolution of crops is highly contentious and complicated. Some might have been domesticated more than once, or spread to secondary centres where further diversification occurred. Moreover, some authorities have proposed alternative schemes (Harlan, 1971). Evans (1993) reviews and comments on the history and recent debate of this fascinating subject.

Given the above caveats, the proposed centres of origin of the main crops are indicated in Fig. 4.1. The main point of this diagram is to suggest that few of the crops, if any, are now restricted to their original centres. For example, Fig. 4.2. shows that the major cereal crops which originated from very different parts of the world, are now grown over large areas that bear little relation to their original habitats. Typically, dispersal through trade or migration, has been followed by selection and breeding to create additional diversity at a succession of new sites. The present distributions of most crop species are therefore more often the result of human migration and colonial history than their place of origin.

The cycle of dispersal is still continuing, accelerating even, aided variously by the internationalization of agricultural research, the intensity of modern commerce and the need for confined human populations to migrate. Maize is still spreading to dry areas in Africa and northwards in Europe, following selection and breeding for quite different traits in each region. The most spectacular changes in the distribution of crops in the present century has been the spread and domination of plantation fruit and beverage crops, such as tea, rubber and oil palm in many parts of Africa, Asia and South America. The detailed and informative entry by Burkhill (1935) on rubber, *Hevea braziliensis*, shows how, from a locally useful, more or less natural, tree in South America as recorded by European explorers in the 17th and 18th centuries, it developed in the late 19th century and became in the 20th century a very profitable, large scale crop in parts of Asia, including Sri Lanka, India and Malaysia. The spread of oil palm in South-east Asia is even more recent.

National and international efforts in conservation

Botanists and plant collectors have for centuries assembled collections of crop species but only in recent times have genetic stocks of the main species been systematically collected and kept together (as 'genebanks' or 'germplasm') in national or international archives (Frankel *et al.*, 1995). The Consultative Group on International

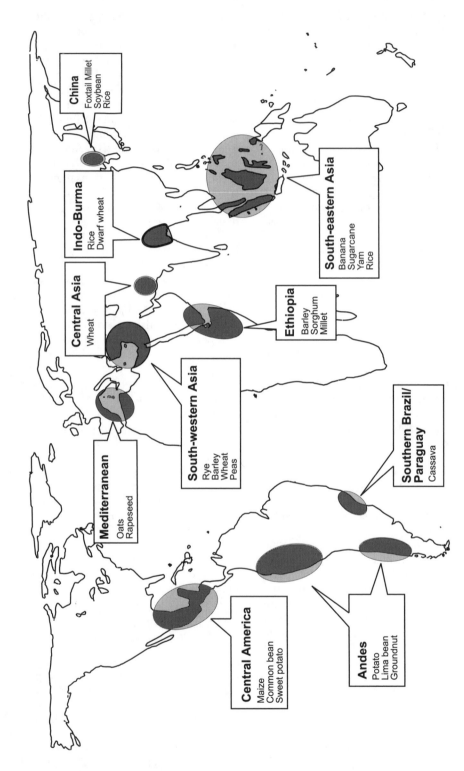

Fig. 4.1. Principal origins and centres of diversity of some for the world's major crops. (From Vavilov, 1928 and other sources.)

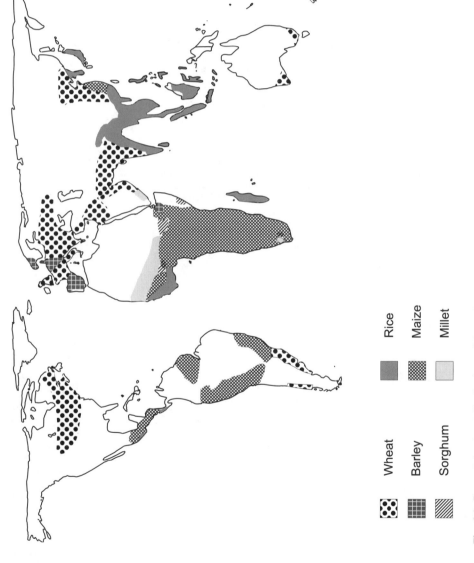

Fig. 4.2. Present dominant distributions of the major cereal crops.

Wheat

Barley

Sorghum

Rice

Maize

Millet

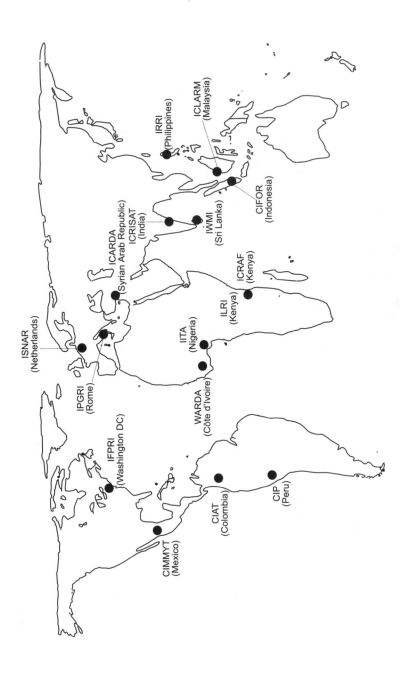

Fig. 4.3. The International Agricultural Research network of the Consultative Group for International Agricultural Research (CGIAR): CIAT – Centro Internacional de Agricultura Tropical; CIFOR – Centre for International Forestry Research; CIMMYT – Centro Internacional de Mejoramiento de Maiz y Trigo; CIP – Centro Internacional de la Papa; ICARDA – International Center for Agricultural Research in the Dry Areas; ICLARM – International Center for Living Aquatic Resources Management; ICRAF – International Centre for Research in Agroforestry; ICRISAT – International Crop Research Institute for the Semi-Arid Tropics; IFPRI – International Food Policy Research Institute; IITA – International Institute for Tropical Agriculture; ILRI – International Livestock Research Institute; IPGRI – International Plant Genetic Resources Institute; IRRI – International Rice Research Institute; ISNAR – International Service for National Agricultural Research; IWMI – International Water Management Institute; WARDA – West Africa Rice Development Association.

Agricultural Research (CGIAR) and its international centres (Fig. 4.3) such as CIMMYT in Mexico, ICRISAT in India, IRRI in the Philippines and CIP in Peru, all have responsibility for collecting and systematizing landraces, wild relatives, commercial varieties and breeders lines for each of the main subsistence species. The International Board for Plant Genetic Resources (IBPGR) was established in 1974 to stimulate and coordinate international collections. On a more local scale, national research organizations have their own collections of seed, tubers or cuttings. The International Plant Genetic Resources Institute (which replaced the IBPGR), based in Rome, now coordinates the efforts to save and explore genetic material relevant to tropical crops.

Genotypic markers and identifiers

The move to systematize and conserve plant material, begun by the classical botanists, relies on there being measurable differences between plants that enable them to be distinguished. Like can be grouped with like, and their histories and relatedness inferred. Appearance or quantitative measures of size and shape are usually enough to distinguish many higher plant species. Chromosome studies among closely related groups of species give indications of their evolution and relatedness. Within species, morphology can still be used to an extent to distinguish ecological or genetic types, particularly if the traits are stable, such as flower structure usually is. Sorghum, for example, was classified into five main races – *bicolor, caudatum, durra, guinea* and *kafir* – based largely on flower and grain morphology (Harlan and de Wet, 1972). Vegetative traits are less reliable, though the extreme forms of tea – the Assam with large, horizontally held leaves, and the China with smaller, more upright leaves (Kingdon-Ward, 1952) – can still be used as preliminary identifiers of some types, though there is such a wide range of leaf morphology in modern clonal tea that the range between Assam and China is more or less continuous. It is much more difficult to detect physiological traits associated with tolerance to abiotic factors, without actually growing the plants experimentally.

As diversity proliferates within species and as interest focuses on narrower genotypic differences, more discriminating methods are being introduced. A range of techniques that identify parts of the plant's DNA are increasingly used to look at within-species differences that are not readily detectable in any other way (Bachmann, 1994; Karp and Edwards, 1997; Ayad *et al.*, 1997). The approach has been applied to estimate the genetic 'distance' between plant germplasm accessions taken from different parts of the world or from different ecological habitats (e.g. Orozco-Castillo *et al.*, 1994). Once the accessions have been collected, the methodology itself is not prohibitively expensive at a national scale and is likely to become standard in laboratories, even in the developing tropics.

Knowledge of markers of any type – morphological, chromosomal, molecular – can only be relevant to physiological agronomy if it is indicative of some trait in the plant that affects its performance in the field or that can be manipulated by management. The more useful markers have tended to be those that are stable, reproducible and control or are linked to one major function. It is not at all certain, for example, that modern molecular techniques will refine the search for stress tolerant crop types. Such techniques may indicate genetic difference but will be valuable to agriculture only if they announce the presence of a useful trait. (We return to the coupling between marker and function in Chapter 8.)

Ecological preferences and the contribution of agronomy

The distribution of a crop species among ecological zones depends first on the range of environments it can tolerate. Genetic manipulation will broaden this range, while agronomy can modify the local environment. Experience has shown that a crop's place of origin is not usually central to defining ecological tolerance or preference, but the climatic conditions at the place of origin are often instructive. To take a simple example, among cereal crops (Fig. 4.2) wheat can grow across a latitudinal range that extends from northern Europe to southern Australia. It is virtually absent at lower latitudes, where high temperature and scarcity of water restrict growth. Where wheat does grow within the tropics, as in subcontinental India or upland Africa, it is either cultivated as a rainfed winter crop or at cooler higher altitudes. So far, the temperature range in which wheat originated and in which it can now grow and survive has precluded its expansion to the hottest parts of the tropics.

A tight sequence of climatic tolerances for different cereals is illustrated on a south-west transect from sub-Saharan Africa to the west coast of the continent (see Fig. 4.2). Across a very narrow range of latitude, northern locations where it is too hot and dry for the cultivation of any cereal give way to production systems for millet, sorghum, maize and rice in that order from north to south. At the northern limit of arable cropping, a wet season of 120–260 days and an annual rainfall of 500–1400 mm restricts crops primarily to annual dryland species such as cowpea, groundnut, sorghum and millet. Near the coast, a rainfall of >2000 mm allows perennial tree crops and cereals such as rice and maize to be grown throughout the year (Kowal and Kassam, 1973). The geographical location of the cereals' place of origin – millet and sorghum from Ethiopia, maize from Central America and rice from South-east Asia – has no significance in determining this sequence. Much more relevant are the physical conditions at these places, and the way they have determined the plants' water requirement and thermal tolerance. It is also instructive that breeding and agronomy have increased the potential overlap between the species on this dryness gradient but have not done so to the extent of eliminating the sequence of species altogether.

There are many other instances where breeding and agronomy have worked on the inherent physiological properties of species to fit them into a particular physical environment. The history of the modern plantation crops demonstrates, perhaps more than any other set of examples, the strong and rapid contribution of agronomy, in particular, in converting a locally useful species into a very profitable international industry. The stimulus for rubber as an industrial commodity came from several developments in processing latex, such as that led by Goodyear in South America (see Burkhill, 1935). However, the means of converting this poorly yielding tree into a major agricultural enterprise in Malaysia and elsewhere in South-east Asia, which began in the late 19th century, were entirely agronomic – better methods of tapping the latex, of propagating and establishing plants, and of devising the optimum stand structure and soil fertility. There were no precedents and very little experience to refer to from the country of origin. A general feature of the initial spread of crops such as rubber and oil palm was the narrow genetic base of the plants used to initiate the new industries. Better management made the initial transformation; genetic improvement came later with selection and breeding of the local stocks or importation of new genetic material.

Agronomy has repeatedly contributed in such ways to the dispersion of all crop species by modifying the local conditions; and indeed, the recent history of crop expansion has given many instances of genotype–husbandry combinations not travelling well together. Again, the plantation industry provides instruction. Tea, for example, has been grown in China for thousands of years, but only came to international prominence through the north Indian plantation industry, from where it has spread in the previous 100 years to south India, Sri Lanka, South-east Asia, central and southern Africa, parts of South America and Australia and the area around the Caspian Sea. Tea now experiences conditions ranging from the hot, humid tropical lowlands to the winter snows of Georgia and Kazakhstan. Its tolerance of such a spread of conditions lies in part in the vegetative character of its yield; reproductive processes would not be so tolerant. Nevertheless, the associated agronomic practices had to be repeatedly appraised as plantations spread to Africa, first in Kenya and later to Tanzania, Malawi, Zimbabwe and South Africa. At first, the agronomy was often uncritically inherited with the genetic material, but had to be altered to suit local conditions before high yields were gained. Shade trees, for example, were first considered essential, but eventually found to restrict yield in the cooler, upland climates of Kenya. Eventually, new clones were developed through empirical selection, much better adapted to the African conditions. They, together with the new agronomy, eroded then obliterated the perceived yield ceiling of the crop.

4.1.2 Selection and genetic manipulation

It is argued that the evolution of crops has occurred through a mixture of natural selection within the cropping environment and directed selection and breeding. Certain traits will have been favoured naturally as a result of using saved seed for the next crop (Donald and Hamblin, 1976; Simmonds, 1979). Genotypes with uniform and more rapid germination, for example, will tend to be encouraged during early evolution of a crop, simply because late seeds or seeds that need a stimulus to break dormancy will contribute fewer seeds to the harvest than those plants that developed from the first flush of emerged seedlings. Retention of seed in the panicle or pod will similarly have been selected independently of any directed breeding.

Systematic, intensive plant breeding is a relatively recent enterprise compared with the timescale of crop evolution, yet has had remarkable effects on agricultural production and the rural scene. As discussed above, the quantitative effects of breeding on yield and profit are difficult to determine, since in most instances new genotypes have been introduced along with new agronomy. In developed agriculture, fertilizer and irrigation might have realized the potential of new cultivars; but conversely, it can be argued that new cultivars were selected only because the fertilizer and irrigation were available. Nevertheless, estimates in Europe

and North America suggest annual yield increases due to breeding of 1–2% in the 20th century (Innes, 1992a).

Mating systems

Plant breeding attempts to increase inherent genetic diversity by sexual hybridization, selects those individuals in the progeny that have traits conferring some advantage or desirable character, then tries to introduce and stabilize those traits in a population. Simmonds (1979) identifies four types of mating system or products of breeding: outbreeding, inbreeding, hybrid, and clone. In sexual reproduction, the primary classification is between cross-fertilization or outbreeding, where fertilization is effected wholly or mainly between plants, and self-fertilization or inbreeding. Outbreeding predominates in plants, and among outbreeding crops are maize, pearl millet, sorghum sweet potato, brassicas and most fruit trees. Many of the outbreeders are still able to self-fertilize, though particularly among the fruit trees, certain mechanisms promote outcrossing. Papaya has separate male and female plants (dioecy) while oil palm produces strings of male and female flowers sequentially over time. Inbreeding through self-pollination is more common in annual crops species, including wheat, rice, barley, groundnut and pea. The progeny of inbreeders will tend to be more uniform.

In some circumstances, breeding purposely generates inbred lines that are uniform in desired traits but the inbreeding also exposes certain harmful traits that reduce survival or yield. To mask these, the lines are crossed or hybridized. The resulting hybrids, having hybrid 'vigour' or heterosis, are usually at least as high yielding as the original populations from which the inbred lines were derived. Hybrids, however, have the disadvantage that they have to be continuously produced from the inbred lines. Generally, farmers cannot save the seed, which must be bought for each sowing.

The fourth system, the clone, is maintained typically by vegetative propagation, in which each individual is potentially the same. Most new tea and rubber plantings are now clonal. There is also an increasing range of cellular and molecular technologies that are able to generate clonal material – from tissue culture for example – in species that are difficult or impossible to propagate from cuttings. Clonal agriculture can be exposed and fragile when dominated by one or a few clones. There are several instances of clonal plantings of perennial crops becoming susceptible to disease after several years' growth. A notion has arisen, therefore, that clonal agriculture must be low in diversity, but it need not be. Given a highly diverse population of mother bushes, clonal propagation can capture and maintain the characters of extreme, rare or valuable individuals that would otherwise be lost when those individuals die.

Forster *et al.* (1997) provide a historical survey of the methods and some of the results of breeding and genetic modification. These authors also consider some of the more specialized techniques, such as combining whole genomes from different sources (as occurred in the evolution of bread wheats), manipulating chromosomes and stimulating mutations. Finally, they consider the modern methods of genetic transformation, whereby genes can be moved between species by means other than normal sexual reproduction, thereby overcoming any natural restriction to crossing.

Major genes and polygenes

Since the early years of plant genetics, it has been known that some traits are controlled very simply by one gene, possibly modified slightly by other genes. The effects of such single genes or major genes can be considerable, completely altering the phenology or biochemical composition (Simmonds, 1979). Perhaps the most well known major gene effects in crops are those giving the short stemmed (or semi-dwarf) varieties of rice in Asia, many of which possess the sd_1 recessive gene (Dalrymple, 1986a) and which, with fertilizer and irrigation, yielded several times more than landraces. Equally spectacular, but more localized on the globe, are the effects of changes in oil quality and content in the seeds of oil palm.

Other major gene effects are the transition in peas from leafy to semi-leafless and leafless and the non-shattering trait of wheat ears.

Most physiological traits, however, are polygenic, determined by several or many genes. It is harder to maintain such traits since it is much more difficult to identify and to keep several genes together so that the desired phenotypic trait is expressed. Resistance or tolerance of water stress, for example, is mediated through many physiological processes. Different processes will operate in different seasons and soils. Accordingly, selection for tolerant traits and their incorporation within breeding lines has been less successful and spectacular.

Both the nature of the breeding system and the extent of polygenic control affect the variability of traits expressed by a genotype. This variability affects yield and performance, as is most apparent in perennial tree plantations, where certain individuals produce great yield while others produce none. A move towards genetic uniformity has been a natural development in intensive agriculture, where there are benefits in each plant producing a similar fraction of yield more or less synchronously. However, there are grounds for considering that heterogeneity in the genotype should itself be valuable in subsistence conditions where the soil and season's pests and weather are themselves inherently variable. Despite its perceived importance, a genotype's variability, and the consequence of this for yield or stability of yield, have seldom been quantified in agronomic study, while crop physiology, following laboratory practice, has tended to consider or demand uniformity from its plant material in order to obtain reproducible responses.

The diversity and uncertainty of crop genotypes

New genotypes are rarely taken up uniformly and exclusively in a region. Rather, a species of crop plant will likely exist in several genetic forms (Hawkes, 1983). Oldest of these, especially of the subsistence crops, will be the landraces, often genetically diverse cultivars that have arisen over possibly thousands of years from the action of natural selection and some breeding by farmers. Traits such as flowering and maturity dates will have been selected and therefore locally adapted so as to fit within important times of the annual seasonal cycle (Donald and Hamblin, 1976). Perhaps existing with or near these will be one or other related wild species. If the mating system is outbreeding, the crop will occasionally outcross, or exchange genes with these species. Then depending on the degree of contact of the site with modern developments in breeding, there will be a proportion of new genetic types, including current and obsolete commercial varieties, and possibly some breeding lines or stocks not yet widely released to farmers. In some areas, the new types will have completely supplanted the landraces. In other areas, the species might only have been brought in and cultivated recently, such that there are no landraces and no agronomic history of that particular crop. Commonly in these circumstances, it is new genotypes, some totally untried, that are the dominant forms; and this is true as much for staple cereals and legumes as for profitable plantation species.

The new varieties are themselves usually of a very complex and sometimes uncertain pedigree. Breeding programmes have often combined genotypes of diverse origin and properties in order to capture a range of disease resistance, yield quality and phenological traits. The well known IR 36 rice has been bred with contributions from 13 varieties found in six countries and a wild species *Oryza nivara*. Dalrymple (1986a) shows the even more complex composition of later IRRI rice varieties such as IR 64, which descends from at least 20 original landraces, variously from China, India, Indonesia, Korea, the Philippines, Thailand, the USA and Vietnam, and a much larger number of parental lines. Breeding for resistance to bacterial blight of cotton has used around 20 genes or polygenic complexes from five species of cotton (Innes, 1992b). This trend to complex genetic backgrounds is a feature of most intensely bred crops. The dominant varieties of the main staples and many other crops are therefore 'world composites', and it is very hard to find any relation between their traits and an ancestral or previous environment. The characterization of such genotypes in relation to the modern environments they might inhabit is therefore one of the most important concerns of agronomy. Again, a preoccupation with ancestral origins is likely to reveal less useful information about the potential of a genotype than an understanding of what physiological traits link it with its environmental niche.

4.1.3 Genotype and environment

It was indicated earlier that many crop species have been found to yield better in some locations and environments that are different from those in which they originated. Conversely, most breeding programmes have provided examples of a phenomenon, which is that a genotype that yields well at the site where it was first bred, fails to yield as well at other sites, and often even at sites that are superficially similar to the first one. Such discrepancies arise because the genotype interacts with many environmental factors in a complex way. The original site was one point on a complex response 'surface' defined by the way the plants responded to radiation, rainfall, temperature, pests and other factors. Yield is only economical at regions of the surface where none of the main factors is seriously limiting. One of the main functions of agronomy is to modify the local environment so as to move it to one of the economical regions of the response surface.

Regression analysis in genotype–environment trials

Traditionally, the economical regions of a response surface are sought through broad-scale genotype–environment trials. The responsiveness of a genotype to different environments is examined by growing a group of genotypes at a range of sites. If there are enough data from a wide range of conditions, graphical plots can be drawn of the

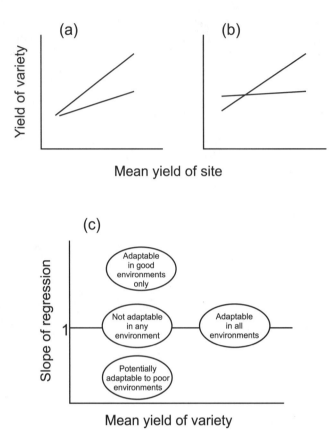

Fig. 4.4. Examples of genotype–environment analysis. The horizontal axis is the mean yield of trialled varieties at sites or defined environments within the trial network; the vertical axis is the yield of a variety, showing (a) two varieties diverging over all sites and (b) crossover, where one variety has a higher yield than the other at high productivity sites but a lower yield at low productivity sites. (c) The main genotypic criteria are mean yield of a variety (across sites) and slope of the regression. (Part (c) after Finlay and Wilkinson, 1963.)

yield of one variety on another, or of the yield of a variety on the mean yield (site mean) of all varieties (Hill, 1975; Simmonds, 1981). The slope of the linear regression of a variety's yield on site mean yield has been used widely to give an indication of mean yield but also the responsiveness of a variety to poor and favourable environments. In Fig. 4.4a and b, for example, lines represent hypothetical varieties with different responsiveness to the range of trial environments; those in Fig. 4.4a diverge as the site mean yield increases, indicating that one variety (the upper line) is more responsive to favourable environments than the other variety. The value of this approach was well demonstrated in the large-scale varietal trials of maize in East Africa in the 1960s, (Eberhart *et al.*, 1973; Darrah and Penny, 1974). These authors condensed and summarize large volumes of data, from which they identified both responsive and stable varieties.

Attention has also been drawn to the limitations of the approach, or more strictly, to the misapprehensions that can arise from interpreting the regression lines uncritically. Both the environmental factors underlying yield differences between sites and the physiological factors causing differences between genotypes have rarely been identified, still less quantified, in such trials. Therefore the regression slopes (whether linear or of another form) have nothing other than relative meaning and are not generally transferable to the most adverse conditions. The reader is referred to Hill (1975) whose approach is highly informative, specially of the historical development of the subject since the early years of genetics, and to Evans (1993, Chapter 4) for a critical discussion of regression analysis in genotype–environment trials.

Stability and crossover

In their analysis of barley varieties in Australia, Finlay and Wilkinson (1963) concluded that the two most useful criteria of a variety's performance are the regression coefficient (the slope of a line in Fig. 4.4a or b) and the mean yield of the variety over all environments. They found a range of slopes and mean yields that fitted into the scheme in Fig. 4.4c. A slope of 1 indicates the variety has approximately average stability, its yield increasing proportionately as mean yield increased at more favourable sites. The ideal, but unobtainable, variety for monocultures would lie to the right of the coordinate space in Fig. 4.4c – high mean yield and stable over environments, even poor ones.

Many modern world composites yield well under most physical environments provided that they are given enough fertilizer and nutrients (Dalrymple, 1986a, b). Breeding has reduced their sensitivity to global factors such as photoperiod and to pests and diseases. They tend to lie therefore between the top left and middle right of Fig. 4.4c. This means that the same, apparently stable, high yielders do not necessarily yield well at infertile, dry sites where there are few inputs. The regression analysis referred to in the previous sections has often indicated that the slopes of the regressions for two varieties cross at some level of input or environmental severity. Above the point of crossover, one variety is superior, below it the other (Fig. 4.4b). More realistically in subsistence agriculture, varieties are sought that lie in the lower middle part of

Fig. 4.4c: they might have low or moderate mean yield but achieve this in most environments.

The existence of yield crossover (or even convergence at poor sites) in variety trials has a strong bearing on breeding genotypes for severe conditions. In recent times, most selections have been made in good conditions where most of the research stations are sited. Experience shows that selection for high yield here does not always transfer to the typical subsistence farm. Perversely, the varieties that yield well in good conditions or with high inputs suffer the greatest percentage reduction of yield in severe conditions. Nageswara Rao *et al.* (1989b) demonstrate this for groundnut cultivars. Those yielding 600 g m^{-2} with irrigation suffered a yield loss of 75% to 150 g m^{-2} with late season drought. Those yielding 300 g m^{-2} with irrigation lost 50% with late drought, again to a yield of 150 g m^{-2}.

The physiological basis of such differential performance in yield has rarely been investigated. Given that the genotype–environment trials are operating within a complex multi-dimensional response surface, there will be many places (i.e. environments) where two genotypes perform similarly, other places where one outperforms the other and vice versa. Trials carried out over a range of environments might or might not cover points of crossover in yield between the genotypes. Whether or where crossover is found probably has limited significance outside the environments examined. The finding of stability across environments is similarly of limited significance unless substantiated by physiological study. An equally general and related point is that selections and trials should be done at poor sites as well as good ones. This might seem self-evident but many agricultural research enterprises are still centralized and need to be balanced by variety trialling on out-stations and farms.

4.1.4 Implications of genetic manipulation for agronomy

Agronomy will be continually faced with

new planting material, whose requirements have to be assessed in relation to existing genotypes and management. Agronomy has to ask how the new material might differ from known material in the way it alters the flows of resources and the type of products it will produce. How much, therefore, of the genetic background needs to be known in order for agronomy to find the optimum conditions quickly and economically? For instance, are the rate and duration of a crop's transpiration influenced by the number of chromosomes or the level of ploidy (sets of chromosomes) and must the optimum population density be different if a genotype has heterosis?

Questions of this type have rarely been studied quantitatively using modern approaches to resource capture and resource use. Nevertheless, agronomy can be helped through the complexity by three features of the evolution of crop plants. The first is that some of the main traits of crop plants have converged during evolution (Donald and Hamblin, 1976). Examples include a trend to larger seed size (to a point), a reduction in plant height against the natural advantage of tall plants in competition, less vegetative branching and extension (again, against a natural competitive advantage) and centralization of yield. In physiological terms, crop species have become more similar. The second is that some genetically based structural changes – larger leaves and fruits related to increased ploidy – have a negligible effect on the functioning of canopies. Stand density or configuration can be modified in a field crop to optimize resource capture and partition. The third is that the genetic background and composition of many crop species have little influence on some important factors in resource use, notably leaf and canopy gas exchange rates.

Accordingly, physiological principles and methods can be transferred among crops and environments and certain features of stands can be anticipated whatever their genetic composition. More useful than genetic background *per se*, therefore, is knowledge of traits defining general architecture, flowering and maturity date, pest resistance, certain aspects of yield quality and degree of responsiveness to inputs. Agronomy can use these traits as a basis for experiments and modelling. The rest of this chapter examines between-species and between-genotype variation in the agronomically important plant traits.

4.2 Characteristics of the main crop types

The final products of gene expression are very different in, for example, a tuber cell of cassava, a fruit cell of coffee and a leaf cell of tea. Many of the tropical crop species can be distinguished by their products alone. Virtually all crop species are identifiable given additional knowledge of plant size and perhaps leaf shape. In a similar way, crop science has also sought to quantify botanical traits that are useful in defining the way that species and groups of species interact with the main resource fluxes. The similarities and differences at these higher orders of classification are useful because cultivars of a given species might have certain common characteristics, irrespective of any other differences between them.

The use of resource flux (Chapters 2 and 3) as a basis of comparison allows species and higher taxa to be compared by common variables. For example, the timing of development in many species responds to temperature in a similar way. Most crop species differ, not in the basic shape of this response, but in the parameters that define the intercept and slope. Similarly, species might differ in the amount of leaf they display and in the way that this is distributed in three dimensions, but the general shape of the relation between leaf area and radiation capture is again similar for groups of most species. Generalizations are possible if consistent similarities or differences can be found.

4.2.1 Plant architecture

Size, structure and other architectural properties cause the main differences in resource capture between, say, a short duration cereal, deep water rice, a coffee bush and a sago palm. All structural attributes ultimately depend on the rates and durations of processes that control extension and strengthening of tissue. While less is known of root architecture (see Van Noordwijk *et al.*, 1996, 1997) than shoot architecture, similar principles apply to both systems. The different combinations of rate, duration and the inherent branching pattern give rise to an extremely wide range of morphological crop types. No one architectural type need be restricted to a specific agricultural environment: many types can be seen together in subsistence gardens of the humid tropics, for example.

The rates of germination and of early extension and growth are often strong discriminants of species. Sorghum leaves, for instance, generally extend faster than cocoa or tea leaves but, in stands where within-species competition is not too strong, these traits do not primarily determine eventual performance. The main and perhaps universal discriminants are in the arrangement and longevity of meristematic tissue. Together they determine the pattern of branching and the opportunities for storage and reproductive growth. Among the many specific traits related to architecture, three are of central importance: the period for which new tissue is produced; the degree of centralization of vegetative and reproductive tissue in one as opposed to many stems; and the capacity for storing carbohydrate and nutrients.

Phase durations

The overall duration of meristematic activity determines whether a crop is a short or long season annual, or a perennial of a few years or 100 years. Several distinct developmental periods can be identified in the life cycle. A purely vegetative phase gives way to a reproductive phase and this in turn to a fruiting phase where reproductive structures fill and mature. In a determinate cereal species, the three phases occur in sequence and hardly overlap on any single culm. In an indeterminate legume or fruit tree, vegetative and reproductive development continue after fruiting has begun.

The sequence of developmental states (the phenology) of a species is particularly important in determining whether it can be grown in a particular environment. Species adapted to short seasons of rainfall, for example, are adapted as much through their phenology as for any inherent ability of their tissues to withstand drought. Two opposing effects determine the optimum phenology for any site. The first depends on leaf or root longevity: plants get a greater return from investment the longer resource-capturing tissue exists. The second depends on length of the life cycle: a shorter cycle gives plants a greater chance of reaching maturity in a seasonal or uncertain environment. These effects are quantified in the next section of this chapter and their implications are considered in Chapter 8.

Plant species also differ in the degree to which phenology is sensitive to environmental factors such as temperature and photoperiod. Such differences are not major discriminants of tropical crop types but do commonly distinguish tropical and temperate species. The threshold temperature is around 0°C for most temperate cereal species and 10°C for many tropical crops. (These are not rigid limits; some species have values in between.) Most tropical crops have broadly similar temperature ranges and – as discussed in the first chapter – this is what most of all makes them tropical crops. For instance, sorghum, groundnut and tea might look very different, but all stop developing if the temperature falls below 10–12°C. Most tropical species are short-day plants in that daylength has to be less than a certain value before they flower. Many temperate species are long-day plants. Some other species and cultivars are not sensitive to daylength or photoperiod. Even if the weather were warm enough, most tropical cereals would not flower in northern Europe or would take a very long time to flower.

Centralization and storage

Centralization of structure in mostly one

stem, or in a main root, occurs in some species or varieties of most crop types, including cereals, fruit trees and palms. It is not, therefore, a good discriminant of taxa. The degree of centralization is unlikely to be the same for shoot and root systems of a species. The coconut and oil palms have a centralized shoot system but a highly distributed root system. Likewise, a single-culmed cereal retains its option to extend many nodal roots. In shorter types such as cereal and legumes, that will have been habitually grazed as wild plants, the ability to branch profusely is retained in many species. Having most foliage and grain on one stem is convenient, and more efficient in channelling dry matter into reproductive yield, but such stems are more prone to fall over, and to suffer total loss after one bite from a grazing animal. Centralization, especially in crops of short stature, is therefore risky in severe conditions and is commonly avoided in subsistence farming where species and cultivars are chosen that provide some means of spreading the risk. Typical traits giving risk-spreading are

asynchronicity in branching (tillering in cereals) or a propensity of the plant to keep several meristems alive during periods when growth stops.

All plants are able to store assimilate and nutrients – seeds and fruits are themselves a primary store – but a small number of tropical crops are grown mainly for vegetative storage products. The physiological purpose of a store is to hold metabolites in a reasonably secure place for use in later growth. For agronomy, vegetative stores have one of two main functions. They are either transient or mobile, for use by the plant later in its current life cycle, or else in effect immobile because they are accumulated then removed when the plant is harvested. Mobile stores exist in most species as distributed stores mainly of carbohydrate or protein in stems and leaves. They can be remobilized and translocated to expanding foliage or grain when current assimilation or uptake do not meet requirements. The immobile stores tend to be concentrated in specific structures, commonly below ground, and it is the presence of these

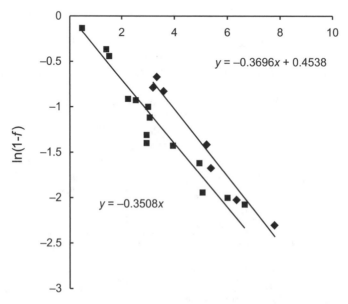

Fig. 4.5. The similar relation between leaf area index and light interception, expressed as ln(1-f) where f is fractional radiation interception, for stands of pearl millet (square) and oil palm (diamond).

(rather than the transient stores) that distinguish species such as potato, yam and cassava.

4.2.2 Resource capture and investment efficiency

The large differences among species in size and structure do not necessarily lead to corresponding differences in the rate of resource capture. By removing environmental limitations, all crop species can achieve dense canopies that intercept most of the solar radiation. Traits such as leaf orientation still influence the radiation extinction coefficient (Chapter 2). Species whose leaves are flat and held horizontally intercept more radiation per unit leaf area than those with leaves held more towards the vertical. Clearly, a broad-leaved canopy such as cocoa will generally intercept more light per unit leaf area than fine-leaved millets, for example, but these differences will be felt mainly during canopy expansion. Once leaf area index (L) rises above three or four, additional growth increments of leaf result in little extra total radiation being intercepted. On any day, therefore, similar amounts of radiation will be captured by groundnut or coconut canopies whether L is 4 or 10.

Among species and higher taxa, traits affecting the attenuation of light by canopies are not strongly discriminating. In some instances even, canopies of very different structure attenuate radiation similarly. Each increment of leaf area index in the canopies of a short-season pearl millet and the massive perennial oil palm intercepts a similar fraction of the incident radiation (Fig. 4.5). They each capture less, per unit L, than a crop with broad leaves held horizontally, the millet because much of its leaf is held at a steep angle, the oil palm because the foliage is concentrated around the centre. The conclusion therefore is that factors such as radiation and carbon dioxide interact with monocultures in a *similar* way whatever their inherent structure and size.

The main differences in the amount of resource captured between crop species is caused by their different longevity. Among a wide range of crop types, leaf longevity and life cycle duration much more strongly determine variation in seasonal or annual resource capture than do maximum canopy size or extinction coefficient. Figure 4.6 shows the relation between life cycle length and the fraction of the annual radiation

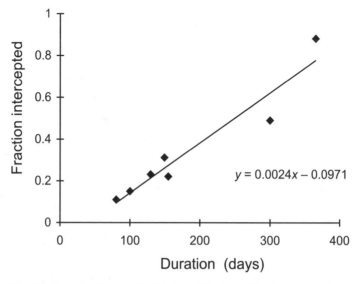

Fig. 4.6. Crop duration and the fraction of the annual total solar radiation intercepted for a range of tropical crops. (After Squire, 1990.)

Table 4.2. Water uptake by cotton, sorghum and maize growing on a single flood irrigation. (After Williams, 1979.)

	Maize	Sorghum	Cotton
Developmental habit	Determinate	Determinate	Indeterminate
Period of root extension (days)	50	80–90	>100
Downward rate of root extension (mm day^{-1})	2.8	3.0	3.4
Maximum depth of water abstraction (m)	1.5	3	>3.5
Period of water abstraction (days)	65	100	115
Total water use (mm)	300	380	520
E_t/E_o[a] at:			
40 days	1.1	0.80	1.2
60 days	0.35	0.6	1.05
90 days	0	0.25	0.6

[a]E_t is crop water use (including soil surface evaporation); E_o is potential evaporation (see Chapter 2).

intercepted by a range of crops (including cereals, legumes, cassava and oil palm) growing in a similar radiation climate. The fraction increases with life cycle duration, by about 0.0024 per day over the range shown, largely because of the decreasing effect of 'lost' time when the canopy was small (see Chapter 5). Similarly, the water extracted from a deep soil increases in proportion to the period of root extension. The example in Table 4.2 shows the greater rooting depth and transpired water by the indeterminate cotton than the determinate maize or sorghum.

The longevity of tissue has an important influence, therefore, on the long-term efficiency of the resource capturing surfaces. We may define this 'investment efficiency' as total (or cumulative) resource captured per unit dry matter in the capturing structures. Returning to a previous example, each unit of leaf lasts for about 60 days in a short-season millet but for over 2 years in the perennial oil palm. Since a similar amount of structural dry matter is required to support a unit leaf area in both species, the oil palm – simply because its leaves live longer – captures more solar radiation per unit leaf area than the millet. The greater longevity of oil palm leaves increases the efficiency three- to fivefold. Table 4.3 compares the investment efficiency for solar radiation in several crops and demonstrates the importance of foliar duration.

4.2.3 Photosynthesis and crop growth rate

The conversion of intercepted radiation during the process of photosynthesis is, generally, less variable among species than are architectural traits, with the exception referred to in Chapter 2 between the C_3 and C_4 groups. The greater efficiency with which the C_4 tropical cereals and grasses use intercepted radiation gives them higher potential maximum rates of dry matter production.

Table 4.3. Investment efficiency of the canopy, expressed on an annual basis as radiation intercepted per unit mass of above ground vegetative tissue, in a range of tropical crops.

Species	Duration (days)	Intercepted radiation (GJ m^{-2})	Investment efficiency (MJ g^{-1})
Pearl millet	68	0.45	1.4
Groundnut	103	0.81	2.1
Pigeon pea	140	1.10	2.2
Cassava	360	4.0	3.1
Oil palm	365	5.5–3.0	3.0–5.6

Monteith (1978) gives a logically argued case that, among stands of the highest productivity, C_4 stands consistently grow faster and produce more dry matter than C_3 stands. Differences between the C_4 and C_3 crops are not always realized, however, and there is some argument as to whether the photosynthetic difference is as important as was commonly supposed (see Evans, 1993). What is quite clear is that possession of the C_4 pathway does not automatically confer greater adaptability or production in hot, dry tropical climates. The C_4 maize, for instance, is often out-produced by the C_3 cowpea or groundnut in dryland agriculture because of other physiological limitations.

Within each of the C_3 and C_4 groups, there is some evidence that stands grown without major limitation of water and nutrients attain a similar conversion coefficient for radiation. Monteith (1977) found that a single value of 1.4 g MJ^{-1} (total radiation) explained much of the variation in dry matter production of a range of species growing in northern Europe. Such values are useful for looking at agricultural production over a large area such as a country or a continent, but should only be transferred with great caution to individual species and crops. The different species in Monteith's analysis each have their characteristic slope. Comparisons of grain legumes in Australia by Muchow (1985) also demonstrate that species differed in the conversion of radiation. Many such comparisons in different environments would be required to define whether differences in the conversion coefficient are really species specific. Estimates in any case face the uncertainty of dry matter partition to roots and the loss of carbon through root turnover and exudation of plant metabolites from roots to soil. Analyses are thereby obscured for woody plants and other perennials with deep rooting systems but also for annuals growing in drying soils, for example, where up to half the measured plant dry matter can be in the extracted root system.

Given these uncertainties, perhaps the expedient approach in agronomy is to work on the following presumptions. For species otherwise similarly adapted to the environment (e.g. sorghum and cowpea, not maize and cowpea), the conversion of intercepted radiation of the C_4 species can be up to twice that of the C_3 but will often be much less than twice and sometimes even similar; while among each of the C_3 and C_4 groups, species-specific differences are probably not large and consistent enough to have as great effects on total production and yield as traits covering architecture and duration.

4.2.4 Source, sink and harvest index

The harvest index (Chapter 2) is not a primary variable in crop production but rather the result of many influences of architecture, total biomass production and sink size. In some species, harvest index is stable over a certain range of environments, to the extent that it has been regarded a constant in some crop models. Plant breeding has moved the harvest index up to around 0.5 in all the main grain and fruit crops, among which the maximum harvest index is therefore not a useful discriminant. The harvest index can be much higher than 0.5 if storage tissue is the economical yield, as in 'root' crops and also those with storage tissue in the trunk, such as the sago palm (Corley, 1986). We return to this theme in Chapter 8.

More important than harvest index itself is knowledge of the functioning of and couplings between the sources and sinks. The source can be taken as the supply of carbohydrate as determined by traits of canopy architecture, duration and photosynthesis. The sink (in agronomic study) is the number and capacity of reproductive or storage structures that are available to receive carbohydrate from the source. Four main features of the source–sink relations of a species impinge on breeding and agronomy. The first is whether the yield is habitually source-limited or sink-limited; the second, the structural plasticity of vegetative and reproductive sinks; the third, the extent and type of coupling between source and sink; and fourth, whether one or other, though usually the sink, is inherently unstable in certain environments.

Is there a main limitation?

The effects of plant improvement in all the main grain (and many fruit) tropical crop species has been to move plants predominantly towards source-limitation, in some instances from an earlier state of sink-limitation. The type of experimental evidence is considered by Evans (1993, Chapter 5) and also Squire (1990, Chapter 5). Generally, source-limitation is found if, in favourable growing conditions, yield declines when the source is reduced, by leaf pruning for example, but yield is little affected by removing some of the sinks (indicating super-abundance of sinks). There are several demonstrations in cereal crops, but one of the most instructive is for oil palm in Fig. 4.7, which shows fruit yield declining as the radiation interception per individual is reduced by pruning and also by planting at high density. Inherent sink-limitation is rarer in improved crop species, but was found in tea, where the yield of shoots is limited more by the low extension rate of buds than by the supply of photosynthate from the substantial canopy (Tanton, 1979; Squire and Callander, 1981). Sink limitation also occurs in some minor crops. The saffron crocus (though not tropical) is an extreme case.

Structural plasticity

The balance between reproductive and vegetative sinks shifts systematically if conditions lead to a reduction of individual plant biomass. In Fig. 4.7, the fruit bunch yield decreases proportionately with reductions in the radiation intercepted, but the extrapolated line intercepts the horizontal axis well above zero. The intercept is evidence that relatively more of the energy is allocated to vegetative tissue than reproductive tissue, thereby reducing harvest index, as the source is suppressed. Physiological mechanisms respond to the lower light to increase the biomass allocated to light capturing and competing organs in the stem and leaf. Plants with great structural plasticity, such as rice and small-grained cereals, often display little evidence of change in the fraction of dry matter allocated to reproductive and vegetative sinks compared with less plastic forms such as the palms.

Such systematic change in the balance of vegetating and reproductive sinks occurs in all plants and can usually be quantified from a graphical plot of reproductive yield on total plant biomass expressed per plant from the data in a plant density experiment. The intercept on the axis for total mass can

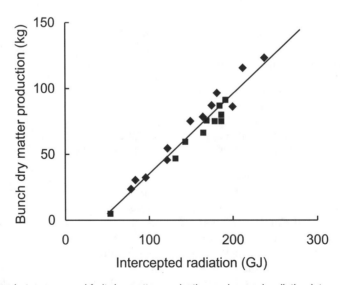

Fig. 4.7. Relation between annual fruit dry matter production and annual radiation interception (PAR) per palm; data from pruning (square) and density (diamond) trials in Malaysia. (Adapted from Squire and Corley, 1987.)

be as low as 1 g in some rice cultivars and as high as 80 kg in oil palm (Squire, 1990). Structural plasticity is a major discriminating trait. Canopies of dryland millets and the palms, for instance, are similar in terms of the light interception per unit leaf area, but the oil palm needs very much more energy to reach reproduction. Both succeed as crops in their own environment. The millets need great plasticity to ensure some yield is achieved in dry years. To counter low plasticity, the oil palm intercepts much more radiation, using its canopy more efficiently in doing so. Forms with low plasticity are only superior when they can intercept and convert substantial resource. Structural plasticity is also one of the main traits that determines the optimum plant density (Chapter 6).

Coupling of source and sink

The sink is so tightly coupled to the source in some species that it is difficult to determine which is driving. In determinate cereals, the resource flux before flowering influences the fraction of grain set (see later in this chapter and Chapter 5). The grain sink is thereby balanced with the store and new assimilate produced during grain filling. Over the life cycle as a whole, source and sink are finely matched. A contrasting characteristic is found in some legumes where

only assimilate produced after flowering is used to fill grain. Here, the sink can only be matched with the part of the source that functions during reproduction.

Less important as a distinguishing trait is the coupling of source to sink. In eco-physiological literature, there is evidence that photosynthesis can be down-regulated (its rate reduced) when there are not enough growth sinks in the plant as a whole to accept the assimilate. Down-regulation is not commonplace in crops since there are usually enough sinks in stem and other vegetative tissue, even if the reproductive ones are removed.

Sink sensitivity

Though most tropical crop genotypes are basically source-limited, they still might suffer sink-limitation under environmental strain. For example, a severe environment or pests might reduce the sink size to a value from which the plant is unable to recover by enlarging individual grains or fruits. Generally, plants are more liable to this form of sink-limitation if the yielding structures are exposed rather then enclosed within protective tissue. It is reasonable to generalize that the sink will be most sensitive in reproductive crops with exposed flowers, such as coffee, and perhaps least sensitive in vegetative crops such as tea. It is less valid

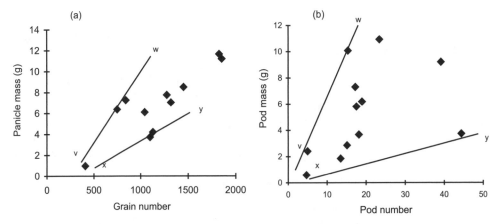

Fig. 4.8. Comparison of reproductive yield per plant and sink size in (a) pearl millet and (b) groundnut in a range of drought and temperature treatments. See text for the meaning of lines vw and yx. (From ODA, 1987.)

to generalize as to which species are less or more sink-sensitive among the broad range of cereals and legumes in tropical agriculture. Several physiological traits, which each tend to endow greater stability, are found to at least some degree in most of the species.

Something can usually be revealed about the sensitivity of the sink itself, from a graphical plot of total grain mass per plant on grain number obtained in a range of treatments. The plots in Fig. 4.8 give contrasting responses for pearl millet and groundnut. The lines **yx** link treatments that produced little grain mass from a large number of grains; in both species, this was because conditions were equable while the sink was being set, but severe drought later reduced the dry matter available to fill the grain (source-limitation). The lines **vw** link treatments where high or low temperature reduced grain number while having no effect on dry matter production, so that individual grains were relatively heavy (sink-limitation). The relation overall was tighter in pearl millet, leading to a more stable individual grain mass. Traits that promoted stability included many small sinks rather than fewer large ones, tight physiological couplings between source and sink, ability to retranslocate stored assimilate and less sensitivity of the sink to extreme temperature. (For further discussion, see example at the end of Chapter 5 and Section 4.3 later in this chapter.)

4.2.5 Functional traits as taxonomic criteria

Tropical crop species tend generally to be identified and grouped by their products, but such grouping is of limited use for agronomy. We have seen above that there are few traits that agronomically define a species in terms of its use of resources and by the way that use might be manipulated. The presence or absence of nitrogen fixation defines the legumes. The type of photosynthetic pathway, C_3 or C_4, is strongly indicative of maximum crop growth rate in good conditions. Otherwise, the species group or type,

such as grain cereal, pulse, tuber or fruit, adds little information pertinent to resource fluxes. Consequently, the label of a species or a higher taxa is not always helpful and can be misleading in agronomy. It is arguably better to dump the baggage associated with such labels and define a stand in terms of the functions that are likely to control its interaction with resources.

There is still potentially a range of functions that can be used to define a stand. They can be reduced to a few main categories, in addition to those for N fixation and photosynthetic pathway:

- life cycle length and leaf longevity, which govern the investment efficiency of resource capturing structures;
- general architecture, specially whether the leaves, roots and reproductive structures are determinate or indeterminate;
- the phenotypic plasticity, defined by the extent to which individual forms might be reduced in size yet still give an economical yield per unit field area;
- source- or sink-limitation to yield and some measure of the additional sensitivity of the sink.

Even qualitative information on these will give clues as to the optimum density, water requirement and yield.

4.3 Genetic modification within species

Here, we concentrate on those attributes that cause large differences among varieties and individuals within species, and which have been most altered by selection and breeding. Simmonds (1979) argues that breeding has had most influence on three main groups of traits. One includes habit, determinacy, perenniality, and plasticity; an important sub-set of this group are traits for developmental timing and longevity. The second determines the biochemical and nutritional composition of structures such as fruits, edible leaves and storage organs, and their associated organoleptic characteristics. The third defines the way plants interact with pathogens and pests. Many of these traits appear to be governed by one or a few genes, though this is less and less so for pest resistance and tolerance. Genetic manipulation has been less successful in altering many other traits, such as those determining the way a genotype reacts to drought, extreme temperature, salinity or waterlogging. Many traits of this type have been difficult to define physiologically, and some operate through traits in the first group above, specially those defining time of development.

Examples of the effects of genetic modification are now discussed, concentrating on those that have had most influence on resource fluxes. (Variation in pest and disease resistance is considered in Chapter 7.) Brief reference is also made to traits that have proven difficult or impossible to change. The maximum rate of photosynthesis per unit leaf area is perhaps the most universal of this class.

4.3.1 Plant architecture and development

In most ecosystems dominated by annual growth of new populations, the main characteristic of the dominant species is their superiority in competition for resources with other plants. Competitiveness is promoted by rapid extension to give large surfaces for resource capture above and below ground. An ability to compete is beneficial where weeds are a major problem, provided individual crop plants are not so close to each other. Given full agrochemical input, however, such competition between individuals of the same crop genotype is no longer beneficial. Increased height growth and greater branching bring in more resources early in life, but then after the canopy intercepts most of the radiation continued competitiveness between crop plants diverts assimilate to vegetative structures and away from yield. In the recent evolution of modern crop genotypes, therefore, the trend has been away from competitiveness (Donald and Hamblin, 1976) and towards less branched, shorter structures. Consequences of this move are that natural competitors have to be controlled, and more seed and more nutrients per unit field area

are required to achieve the high density stands that maximize resource capture.

Most species of the common crop types such as cereals, legumes and root crops already have a range of morphological types. Architecture is particularly well demonstrated by the range of varietal type in rice. *Oryza sativa* L. is grown in many environments, including lowland inundated (52% of total area), lowland rain watered, 'upland' on well drained soils (13%), deepwater (8%) and tidal (3.4%) (Dalrymple, 1986a). Each environment requires very different characteristics. In addition, there are two geographical races. *Indica* is the major form in humid regions of South-east Asia and China. It is adapted to low fertility and uncertain watering, competes well with weeds, resists tropical diseases, and has well-liked cooking and textural features. Its tall, late maturing habit, however, makes stands liable to fall over when heavily fertilized and irrigated. The other race, *japonica*, is grown more in temperate regions, such as parts of China and Japan. It is shorter, matures earlier, is more sensitive than *Indica* to fertilizer and water control but also to viral diseases of the tropics.

Most new varieties of rice are shorter, mature earlier and lodge less easily than traditional varieties of either race. Part of their shortness is caused by their being less sensitive to photoperiod which in sensitive types delays flowering and increases leaf number on the main stem. Shorter types have been known since the early 1800s, but most recent semi-dwarfs have developed from breeding programmes begun much more recently (see earlier, this chapter). Most other tropical staple and subsistence crops have a wide range of architecture, though perhaps not so wide as rice. A single species such as sorghum includes more or less uniculm types to ones that tiller profusely, and a spread of panicle characteristics from very tight cylinders to lax, open, branched structures.

Vegetative and reproductive durations

One of the most important physiological factors to be modified systematically in many species is the time to flowering, or rather the times spent in vegetative and reproductive phases of the life cycle. This is so for perennial fruit crops as much as for the main cereals and legumes. Since vegetative meristems are partially or totally suppressed when flowering starts, any tendency to early flowering also tends to reduce branchiness and indeterminate characteristics. Increased control of the relevant structures is thereby gained. The most spectacular consequence of this trend has been the development of short stemmed, early flowering, green-revolution cereals that yield very highly when given plenty of fertilizer and water. Other examples include the breeding of leafless and semi-leafless peas, which give light penetration deeper in the canopy, better structural stability and less vegetative matter than leafy types (for further examples, see Donald and Hamblin, 1976). The manipulation of canopy architecture to allow greater penetration of light into the canopy results in an adjustment, i.e. a lowering in the value of the extinction coefficient (k) of a particular variety within the usual range for its class of species (see Chapter 2, Table 2.1).

In a broader context that includes subsistence agriculture, earliness is not the only or even the main requirement in developmental timing. Matching phenology with the annual cycles of rainfall and nutrient availability are more important. Earliness is still required in some locations either to avoid a period of scarcity in resource or to permit more than one crop in a year. In other places, an extended phenology is needed to avoid damage by pests or survive periods of scarcity.

The very wide range in phenology of subsistence crops such as sorghum, pearl millet and pigeon pea testifies to the success of genetic manipulation, which has been helped by the relatively simple genetic control of phenology in these and many other annual species. As just discussed, however, developmental timing is itself rarely independent of other traits, notably height and branching. In a given temperature and photoperiod, longer development is associated with more leaf primordia, internodes and branches. Together, these give a more com-

petitive plant, which – as also indicated above – has beneficial and detrimental implications. In this way, phenology is tightly coupled to the degree of weed suppression by the crop and the population density for maximum partition to reproductive organs. At each site, therefore, the most suitable variety will be the one that optimizes the often conflicting needs of (i) matching seasonality of resource, (ii) weed suppression, (iii) seed availability and (iv) population density for the most favourable allocation to yield. The links between phenology and season are considered further in Chapter 5.

Canopy duration and resource capture

The type of response seen between species in Fig. 4.6 is also apparent between cultivars of a species. Since intercepting surfaces are small initially, and commonly only reach their full area or length just before flowering in short-season types, any prolonging of a full surface has a disproportionate effect on total resource intercepted. Figure 4.9a shows this effect from the study by Hughes *et al.* (1981) on pigeon pea in Trinidad. The additional yield of medium- and late-flowering cultivars is proportionately much greater than the extra time itself.

The authors repeated the experiment at four sites, and although the ranking of the cultivars was the same, the absolute values of dry matter, loss of plant material after flowering and total radiation interception differed, showing other site-variable factors were influencing plant production. The arrowed coordinates in Fig. 4.9a show stands where pollination was considered to be unsuccessful. When grain yield is plotted against total biomass (Fig. 4.9b), these five stands lie well to the right of the rest. The remaining stands form a consistent relation, suggesting yield of the early variety was source limited (cf. Fig. 4.7). Apart from pollination, no other traits affected the result more so than developmental timing. Few other studies have compared cultivars over several sites in terms of resource capture but several examples of detailed growth analysis show the importance of an extended

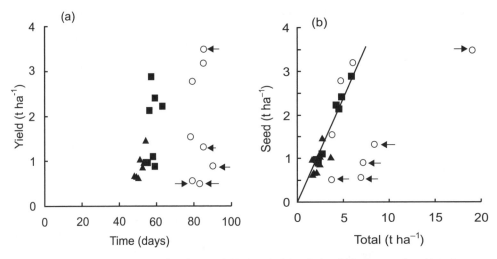

Fig. 4.9. Effect of developmental duration on yield of early (triangles), middle (squares) and late (open circles) pigeon pea in Trinidad, showing (a) grain yield and time to flowering, (b) grain yield and total dry matter. The arrowed points indicate stands having poor pollination; the regression in (b) through non-circled points has a slope of 0.476, forced through origin ($r^2 = 0.88$, $n = 19$). (Original data of Hughes *et al.*, 1981.)

period of high leaf area before flowering (Goldsworthy, 1970).

4.3.2 Quality of product

The traits of interest here are those that define the shape and size, the surface features, the hardness, the nutritional and biochemical composition and general organoleptic features of the particular plant part harvested as yield. Product quality is important for various reasons, including ease of harvesting, identification of the product, post-harvest storability, resistance to pests, taste, food safety and nutrition, cookability, marketability and industrial specifications.

Some important quality traits have been systematically altered by selection and breeding in a wide range of plant families during domestication. For example, mean grain size in many species has increased during their domestication from wild relatives (Evans and Dunstone, 1970). One of the advantages of increasing grain mass is a larger internal volume compared with covering material. One disadvantage for farm-saved seed is that a larger fraction of the yield will be diverted to sow at a given frequency the next year. Evolution during domestication will have resulted in some optimum size to best satisfy local requirements for ease of handling and cooking and having enough propagules to sow the next crop. More recently, the tendency towards specialized seed production systems in high-input agriculture has relieved the necessity for all farmers to keep back seed. Many other traits have been altered more or less in isolation from work in other species. Examples include those defining taste in beverage crops.

The genetic control of some quality traits is simple. Occasionally, a single cross has produced a new type whose adoption by farmers or industry has had great consequences for the productivity of a system or the wealth of a country. The formation of the *tenera* type of oil palm fruit, with its greater proportion of flesh and oil to nut, from the *dura* and *pisifera* types, led to massive increases in the oil yield of the crop. There was no change in total plant productivity, just a switch in partition between oil and other plant constituents, mainly foliage and trunk. This was one of the reasons for the spread of the palm to dominate the land use of South-east Asia. Many other traits are under much more complex control, however.

Stability over environment

There is such a wide range of traits and forms of genetic control that agronomic approaches to quality can be very specialized. Local growers have their individual approaches to a product from which they can get most sustenance or profit. Moreover, principles in this area of study have been slow to emerge because of the dearth of systematic experimentation across a range of conditions. Perhaps the safest first step is to assess the stability of the trait in question from current knowledge or measurements over a range of environments whose physical characteristics cause large differences in total plant mass or production.

Of particular value in agronomy are traits that are stable in the face of change in environment and agronomic input. Stable or conservative traits include not only descriptors of colour, texture and the general nature of the main biochemical product, but also certain features of composition. Examples occur in both high input and subsistence cropping. In oil palm, for example, the oil as a fraction of fruit dry mass is stable over a wide range of total plant production caused by variation in climate and agronomic input. Equally stable in the grain of a high protein subsistence crop such as bambara groundnut are the proportions of carbohydrate, protein and fat. In one of the few systematic investigations of physical environment on product quality, the proximal composition of bambara groundnut was also shown by experiment to be virtually constant over a range of temperature (Brough and Azam-Ali, 1992). It is not clear how such conservatism in the proportions of constituents is regulated genetically, and why it is only weakly sensitive to environment. A stable allometry in the components of individual reproductive propagules is generally achieved at the expense of variability in number of propagules per plant.

Somewhat less stable than percentage composition is the mean mass of grain or fruit. As Fig. 4.8 shows, small grains are produced when the assimilate source during grain filling is much smaller than the grain sink. Agronomy can act to correct such unstable behaviour by balancing the rates of assimilation during grain set and grain filling; examples are given in Chapter 5 on the temporal regulation of source and sink in crop plants. To a degree, the percentage composition of nitrogen is open to influence and is considered in Chapter 7.

Many other traits are inherently unstable. The blend of leaf chemicals, including flavanols, theaflavins, thearubigins, phenolic acids and volatiles, that define colour, taste and smell in tea, are very sensitive to growing conditions (Millin, 1987). One generalization with such traits, though tentative, lies in the common observation that environmental conditions giving rise to rapid growth are rarely those that support the highest quality of product. Growers of the beverage crops and grapevines know by experience that the slowest growing plantations, sometimes on poorer soils, and at high altitude, often produce a product of the highest quality. Individual fields, characterized by their aspect, soil and microclimate, and sometimes landrace, each give a characteristic product. Clearly, the metabolic pathways giving rise to the array of leaf and fruit chemicals are very sensitive to environment. The control of these pathways in the context of resource fluxes requires a generic and systematic study.

4.3.3 Gas exchange

Evans (1993, Chapter 5) has summarized work showing variation in photosynthetic rates among cultivars and varieties of crop species. He draws attention to several important facts.

Genetic vs. environmental control of photosynthesis

One is that the maximum rate of photosynthesis per unit leaf area has not been increased by plant selection and breeding. Indeed, a weight of evidence shows that modern varieties resulting from intensive selection have lower rates of maximum photosynthesis than their wild relatives or landraces. A second fact is that of the *photosynthetic paradox* – that photosynthesis rate per unit leaf area and crop growth rate have not been increased by genetic manipulation, but are routinely increased by improving mineral nutrition or water availability or by raising the carbon dioxide concentration.

To explain these facts, any genetic change in photosynthesis rate must be compensated by a change in the opposite direction in another factor contributing to crop growth rate. One of the most important factors discussed by Evans is that individual leaf area and photosynthesis per unit area are not independent but seem to be negatively related among genotypes. For example, an attempt to breed for genotypes with larger leaves (probably the result of more rapid expansion) seems to result in them having less mass of photosynthetic apparatus per unit area and vice versa. This negative coupling is not normal, however, when environmental limitations are overcome. Adding nitrogen to a deficient soil commonly increases both leaf area and photosynthetic rate per unit leaf area. At the scale of the stand, these magnify to enhance leaf area index, cumulative radiation interception, the conversion coefficient for radiation and crop growth rate (see Chapter 7).

While the basis of the paradox is complex and not clear in all circumstances, its existence helps to simplify matters for physiological agronomy. Genotypes of a species might be expected to differ to a small degree in photosynthesis rate but, in most tropical environments, the greatest scope for improvement in photosynthesis will be agronomic. It is even more important to realize that, while photosynthesis rate is much researched and written about, other traits will commonly be stronger determinants of yield in the field. Perhaps chief among these is canopy duration as related to plant development (see Chapter 5).

Carbon isotope discrimination

Genetic differences in photosynthesis rate or the photosynthesis/transpiration ratio are generally much smaller therefore than those in developmental timing or habit. Reviewers such as Ludlow and Muchow (1990), rank the photosynthesis/transpiration ratio well below phenology, osmotic adjustment and rooting characteristics among traits that need to be manipulated to optimize yield of grain sorghum in the dry tropics. Nevertheless, any increase in the photosynthesis/transpiration ratio would still likely bring benefits. The difficulty of measuring it (and so selecting for it) in the field is that estimates of dry matter production and transpiration are subject to large errors. Particularly, the amount of carbon allocated to fine roots or lost from roots to the soil is very variable in dry conditions and hard to measure accurately.

Techniques involving carbon isotope discrimination have been developed in attempts to get round these difficulties. The abundances of the naturally occurring stable isotopes of carbon (^{13}C and ^{12}C) change during photosynthesis since the processes discriminate against the heavier isotope. Compared to the air, the products of photosynthesis consist of relatively less ^{13}C and relatively more ^{12}C. Moreover, the degree of discrimination has been shown to be related to the photosynthesis/transpiration ratio in C_3 plants: experimental results have confirmed that genotypes with a high photosynthesis/transpiration ratio have proportionately less ^{13}C in their tissues.

The methodology is efficient because the isotope abundance can be determined on a small sample of tissue taken from a field or laboratory experiment. The technique's value in selection and breeding will depend on several additional factors: whether any difference in discrimination is detected between genotypes; whether the ranking among genotypes (in terms of discrimination) is consistent between sites and years, particularly in harsh environments; whether the degree of discrimination is heritable; and whether a particular value of discrimination leads to or is associated with more efficient use of water or greater yield per unit water available. The studies of Hall *et al.* (1994) on cowpea genotypes in Senegal and White *et al.* (1994) on common bean in Colombia are examples of the use of the technique to assess genotypes and develop selection criteria in realistic agricultural conditions. Although they confirmed that genotypes differed in isotope discrimination, they demonstrate some of the difficulties of using discrimination as a selection criterion. For cowpea in Senegal, for instance, some well adapted, productive genotypes had isotope discrimination values suggesting less efficient use of water in the assimilation process.

4.3.4 Drought and other abiotic strains

There has been limited progress in genetic manipulation of physiological mechanisms underlying avoidance or tolerance of drought, high temperature, and other environmental strains. Many traits of a plant's physiology combine to determine its strain avoidance or tolerance, which tend therefore to be under polygenic control, unlike many of the traits referred to earlier in this chapter. Moreover, the genes that control it are not necessarily genetically linked, which makes them more difficult to assemble in one genetic line.

Traits that allow plants to avoid strains are commonly related to developmental timing and habit. Such traits are not primarily responsive to the environmental strain itself, so their benefit or otherwise will depend much on the season. It was shown earlier, for example, that the change to a reproductive state in dryland cereals caused roots to stop extending, water uptake to cease shortly after and generalized symptoms of drought stress to develop within a week or two. Earliness and centralization of structures will be advantageous when there is a short rain season or an otherwise finite amount of water, but possibly disadvantageous if there is a longer season of intermittent rain. In the latter circumstances, decentralization and asynchrony could be more advantageous.

Though timing and habit will generally be important traits, others conferring toler-

ance to water deficit or high temperature (for example) in specific tissues, are crucial in many circumstances (Ludlow and Muchow, 1990). The most sensitive periods for many species are during emergence and early seedling establishment and during flowering. Individuals or varieties within a species might differ in a wide range of characteristics such as leaf movements and leaf rolling which lessen the radiation load on leaves, continued function and lack of damage by high tissue temperature, and increased osmotic potential of tissue so as to withstand lack of water in the environment.

It has already been noted that selecting for yield as such is unlikely to be successful, since it will depend on the particular season and site. An alternative is to select for certain individual traits that are likely to benefit a genotype in many environments and that have no or few negative or yield-limiting effects. Possible examples of such traits are heat tolerance during emergence and early leaf growth and osmotic adjustment, in which plants accumulate solutes which enable them to maintain their cell turgor when the environment is dry (Morgan *et al.*, 1986). The steps that have to be achieved are then: (i) find a method that can be used to quantify the trait among genotypes; (ii) assess the way the trait is inherited (simply or polygenically); (iii) find a means of identifying whether the trait is present and expressed in a different genome; and (iv) find a method of incorporating the trait into new genomes as required.

None of these steps is straightforward. For example, screening sorghum and pearl millet for heat tolerance during emergence and early seedling growth in the field (Peacock *et al.*, 1990, 1993) was done rapidly and with many genotypes, but the inheritance of the trait is complicated. Screening for osmotic adjustment in sorghum was more technically demanding but its inheritance was determined more simply by two major genes (Ludlow and Muchow, 1990). In both these instances, progress might be possible by using DNA technology to identify parts of the genome that are linked to these traits. If a very high association is found between expression of the trait and the presence of the respective genetic material, its presence can be confirmed in new genomes produced by plant breeding.

4.3.5 Source and sink

In general, the same considerations and conclusions will apply to variation within a species as to variation between species. By experiment and observation, it is feasible to deduce whether cultivars differ in being predominantly source or sink-limited. Following the earlier argument, most modern varieties will be source-limited, and selection and breeding will likely be acting to move any sink-limited lines into source-limitation.

The greatest fund of within-species variation occurs in the plasticity of vegetative and reproductive structures and in the sensitivity of the sink to extreme events. Pigeon pea, groundnut, cassava and sorghum, to name four common species, show a wide range of slope and intercept in plots such as that in Fig. 4.8. An example of the effect of enforced sink-limitation, by poor pollination of generally source-limited varieties, is shown by the arrowed points in Fig. 4.9. In most instances, fairly simple manipulation or else a population density trial will demonstrate the main differences in source and sink among a group of varieties.

4.3.6 Untangling the complexity of coupled plant processes

Plant development and growth need always to be studied holistically. Introducing a new or changed trait at one stage in the life cycle will have ramifications for many subsequent stages. Specifically, the mechanisms that govern the components of grain yield often have their basis in what is happening in the plant much earlier. The preceding sequence of events must therefore be considered when evaluating any particular reproductive trait, such as grain number per unit area or mean grain weight, and deciding whether the

variety might be limited by assimilate source or reproductive sink. The following examples of 'heterosis' in rice, of varietal differences in maize and germplasm screening in sorghum illustrate these points.

Hybrid rice

Hybrid rice, developed in China in the 1970s, may give a yield increase of about 1.2 times the best semi-dwarf genotypes. New seed is required each year but its production is expensive and depends on an off-farm infrastructure. Hybrid rice nevertheless presented an opportunity to break through the yield plateau of the semi-dwarf varieties. The mechanism of heterosis is not entirely clear but involves linkage and coupling within the plant. Heterosis manifests itself finally in more spikelets and a higher total grain weight but these features themselves are more likely an effect rather than the cause of heterosis.

The hybrid's advantages start early with higher enzyme activity and faster breakdown of starch during germination. There follows more rapid expansion of seedlings, a greater vegetative surface, more tillers and a greater spikelet number. Rates of photosynthesis or crop growth of the complete canopy seem not themselves affected. The advantage arises because hybrid plants develop and expand faster in the vegetative stage. They intercept more radiation per unit developmental time and thereby gain a capacity to initiate and maintain larger sinks (number of tillers, number of spikelets). The primary consequence of heterosis is therefore an increase in the capacity of the source before flowering; the secondary consequence is an increase in the capacity of the reproductive sink through spikelet number (by the coupling described earlier).

Maize in Mexico

A trial of five maize varieties at El Batan in Mexico (Goldsworthy and Colegrove, 1974) demonstrates the difficulty of assigning a particular physiological cause to a difference in yield. Two extreme and one intermediate of their varieties are chosen here to illustrate the point (Table 4.4). The most

Table 4.4. Population density, grain yield and dry matter in three varieties of maize grown in Mexico. (From Goldsworthy and Colegrove, 1974.)

Variety	Hidalgo 8	CIPA	Zacatecas 58
Population (m^{-2})	9.0	10.3	9.7
Time to silking (days)	95	81	65
Time to maturity (days)	179	162	140
Grains (g m^{-2})	3040	2350	1900
Grain mass (g m^{-2})	760	704	530
Maximum leaf area index (near silking)	7.5	6.5	3.2
Dry matter at silking (g m^{-2})	1330	1120	700
Maximum total dry matter (g m^{-2})	2520	2100	1540

productive variety yielded 1.5 times the least productive. Resource capture was not measured – the experiment was conducted before equipment for measuring radiation became routinely used in crops. However, the sequence of events leading to the yield difference is probably as follows. The earliest variety, *Zacatecas 58*, produced silks after 66 days and matured after 140 days. The latest variety, *Hidalgo 8 × Mexico Group 10*, took 30 days longer to silk and 40 days longer to mature. Traits such as the rate of leaf production and leaf expansion were not measured but are unlikely to have differed substantially. The cumulative and maximum leaf area index (L) depended therefore mainly on the vegetative phase, which was much longer in the later variety. At the time of maximum L, near silking, the difference in L was more than a factor of 2; but this factor in leaf area would have led to a smaller difference in the fraction of light intercepted at any time by the canopy (see Chapter 2). Nevertheless, the combination of larger L and a longer duration of L allowed the later variety to accumulate 1.6 times more dry matter at silking than the early variety. Consistently, therefore, the later variety produced 1.7 times more grains than the early but its grains were slightly lighter, reducing the factor for total grain mass on the main culm to 1.5 as indicated above.

The authors concluded that all the varieties were sink-limited (did not have enough grain sites), specially when compared to a hybrid grown in Zimbabwe, which had a faster grain growth rate and higher grain yield (Allison, 1969). There is an alternative explanation, however, if the sink was cou-pled to the size of the source before flowering. Since that experiment in Mexico, Hawkins and Cooper (1981) showed grain number in Kenyan maize was proportional to the rate of dry matter production, per unit developmental time, before flowering. What might appear to be a sink-limited yield could in fact have been source-limited. In the Mexican case, the greater source-limitation of the earlier varieties would have arisen from the shorter canopy development leading to a smaller photosynthesizing area before flowering. Their smaller sink would have been a consequence of the smaller source before flowering and not some innate sink-limitation.

Retrospectively unravelling source- and sink-limitations among such varieties is problematic without additional knowledge of climatic data and the seasonal course of resource capture. The weight of subsequent evidence points towards the conclusion that the differences in the yield of the varieties in the Mexican trial were caused by differences in the degree of source-limitation. It is still possible, nevertheless, that the Mexican varieties as a group were inherently more limited than the stand referred to in Zimbabwe. Only a quantitative comparison of radiation capture by the canopies in relation to rate of development would elucidate the nature of the limitation between the varieties and sites.

Screening and selecting sorghum lines for drought and heat tolerance

The complexities of strain tolerance are now demonstrated by the efforts to define and identify genetic characteristics for tolerance to drought and high temperature in sorghum. In the 1980s, the genetic base for this programme consisted of over 20,000 collections of seed and breeding lines held at the International Crops Research Institute for the Semi-Arid Tropics. Several hundred of these were screened visually and systematically for drought tolerance (Peacock *et al.*, 1988). Only five of these were subjected to detailed physiological study in a field experiment (Matthews *et al.*, 1990a, b). These five came from diverse backgrounds in China, Yemen, Kenya, Egypt and Zimbabwe, and included all four combinations of early and late maturity with observed resistance and susceptibility to leaf scorching.

The main lesson from the study was that no one genotype was exclusively adapted to the conditions: the relative performance of each changed during the season and particularly before and after a mid-season dry period. Types that later proved susceptible to drought expanded more rapidly into the soil and extracted more water. They did better than the resistant types when growing on a store of water, i.e. in conditions that would normally be considered to be of substantial water strain. However, they were less able to tolerate the even more severe strain that developed when the water began to run out. Then the tolerant types were better able to maintain green leaf, which they later used to produce more growth when conditions improved. As previously summarized by Squire (1990), this advantage of the tolerant types was based in several traits: slower leaf expansion leading to a small leaf/root ratio (i.e. more root supplying a given area of leaf), a correspondingly faster transpiration rate per unit leaf area and much greater leaf rolling which decreased the direct radiation load.

These results with sorghum are typical in that the genotypes were discriminated by several traits. Moreover, the importance of each trait depended on the context, for example the depth and water-holding capacity of the soil and the length of any within-season dry period. No one genotype will be exclusively adapted to a set of general conditions that we might describe as *drought*.

Particular combinations of traits are therefore required to maximize yield in different environments. As indicated earlier, plant breeding has a difficult task to assemble these traits since they are unlikely to be linked and each might be controlled polygenically.

4.4 Conclusions

New cultivars from diverse genetic backgrounds and geographical origins are being dispersed around the world. Landraces are becoming extinct, others are being rediscovered. The fertility of some soils is declining, while other soils are being opened for agriculture for the first time. Weather patterns can shift for 10–20 years and long-term climatic change is a prospect. In this highly dynamic and uncertain scene, agronomy has two main functions that pertain to genotype. It must accept new genotypes, compare them with what is already available and devise methods of growing them that maximize the flux of resource to the economic product; and it must design crop ideotypes (Donald, 1968; Donald and Hamblin, 1976) for deployment at local scales and present them in a form that gives breeders clear and realistically attainable aims.

To succeed in either function, agronomy requires knowledge of both the seasonal distribution of resources and a set of physiologically based descriptors that define in simple terms the way a crop captures and uses the resources over time. The sciences of environmental physics and crop physiology have now progressed to the point where such a set of descriptors can be defined for most tropical environments. Potential dry matter production and yield for a given seasonal distribution of resource can be estimated with much the same facility that meteorologists estimate the seasonal course of potential evaporation. There will be uncertainty in dry environments over seedling survival, the extraction of water from soil and over the allocation of dry matter to yield, but these should not deter responsible calculations of a feasible yield or its likely range. Complex computer simulations are not necessary for the task. Simple back-of-envelope estimates are usually adequate in the first instance (Chapters 2 and 3). The following information on genotypes would allow a first estimate in most environments.

- The seasonal course of development, as affected by temperature and photoperiod, and the plant growth form. From these alone it should be possible to draw – by reference to published work – a curve of fractional radiation interception over the season.
- In dry environments, root habit and the timing of root expansion in relation to soil depth (if in doubt, dig up the root system), together with a rule of thumb indication of soil water availability, which has probably already been measured for soil of a similar type, seasonal water use can be estimated.
- An approximation of dry matter production from the resource extracted, using either the appropriate conversion ratio or information from a similar stand in the literature.
- The maximum values and likely stability of allometric relations among the main organs. Partition fractions will likely be the weak link, except when only radiation is limiting production. Traits conferring stability include a capacity to retranslocate stored assimilate and osmotic adjustment.

Most likely, insufficient and uncertain information will accompany a new genotype, however. It will be known whether the plant has the C_3 or C_4 photosynthetic pathway or whether it can fix atmospheric nitrogen. Information on developmental timing, height, the chemical or organoleptic quality

of the yield and pest and disease resistance are also likely to be available, since these are the traits that were probably altered by the breeding process. More detailed information on the ecophysiology of the cultivar will rarely be available. As considered earlier in this chapter, the gross structure of the genome or its site of origin is not likely to be an influence on its physiology. However, the breeder, the seed supplier and the scientific literature all might provide additional information on the main physiological characteristics of the genotype's relatives. Caution must be exercised since not all this information, especially that on expansion and growth in controlled environments, might be directly transferable to the field or to new sites.

Uncertainty will arise, however, not just from knowing parameter values imprecisely, but by treating the crop population as entirely uniform, which is the general practice in most studies of agronomy and crop physiology. It has been argued that more variability and plasticity in a population endow it with a greater chance of yielding at least something in any of a wide range of adverse conditions. The matters of variability, stability or reliability are not pertinent just to subsistence agriculture. Harvesters of oil palm plantations know which individual plants habitually yield well, and which are stunted or whose vigorous vegetative growth leaves nothing for fruit yield. Tea pluckers know the individual bushes that flush early and rapidly and those whose buds go dormant. Understanding such variation within populations needs a more individualistic approach to resource use and to genotype and is taken up here in the final chapter.

5

Timing (When?)

Decisions on when to sow or plant, when to fertilize and when to harvest are among the most important in agriculture. Managing timing in crops requires some understanding of the seasonal availability of resources and more particular knowledge of the way a plant's developmental phases are influenced by environmental signals, especially temperature and photoperiod. Management has been aided in this task by the wide range of genotypes that has become available as a result of the systematic breeding, conservation and distribution of the main crop species.

Among both annuals and perennials, the length of the pre-reproductive phase is the period that has been most strongly affected by selection and breeding. The introduction of genes for early flowering has generated short-season, short-stemmed cultivars from traditional, tall, long-season types in all the main cereal and legume species. Shortening vegetative development in this way has given managers more scope to match a crop's phenology with the local weather. However, the new genotypes require different, often more intensive management to realize their potential.

This chapter presents general aspects of developmental timing in crops, investigates the environmental control of development, with illustrations of the effects of altitude and latitude, and then discusses the options available to optimize timing in relation to available resources.

5.1 Characteristics of timing in crops

A crop plant progresses through a series of states between germination and harvest. Some of these are visible on the intact plant, or after dissection, and are defined by the presence, or subsequently the number, of leaves, buds, branches, inflorescences and fruits. Important states in a typical cereal are emergence, initiation of the floral apex, flowering, grain set and grain maturity. Other plant processes controlling change in surface area, mass and chemical composition are linked to and influenced by these changes of state. Therefore, knowledge of what controls the transition between developmental states gives insight to a wide range of factors that determine yield and the quality of product. The progression through states, especially when viewed in relation to the seasons, is termed the plant's 'phenology'.

5.1.1 Change of state and change of quantity

Development is here used in a broad and flexible sense to mean that a change has occurred in the type, number, structure, mass or composition of a plant's organs. Developmental periods are quantified by the time (t) that elapses between any two states defined in terms of these characters. Typical developmental periods are sowing to emergence, emergence to floral initiation,

© CAB *International* 2002. *Principles of Tropical Agronomy*
S.N Azam-Ali and G.R. Squire

initiation to flowering and grain set to grain maturity (Fig. 5.1a). Any other pair of states can be chosen to suit particular analytical purposes. The time between the appearance of two leaves or two reproductive sites (e.g. spikelets) is also a developmental period. As defined in Chapter 2 for germination, this developmental time, t, is usefully expressed as a developmental rate $(1/t)$ for examining responses to temperature and photoperiod. Some authors prefer to name this quantity, $1/t$, as a *rate of progress* towards the defined state.

During any developmental period, the changes in number, area and mass can also be expressed as a rate, for example, millimetres per hour of stem extension, number of leaves per day or increase in dry weight of grams per plant per day. These changes can be regarded as changes in quantity, or 'growth' which are different from the all-or-nothing changes in state that define development. This distinction between state and quantity is useful in practice, since the processes controlling change in developmental state are generally more stable than those controlling change in quantity. For instance, the period between emergence and floral initiation in tropical cereals is often stable in the face of change in population

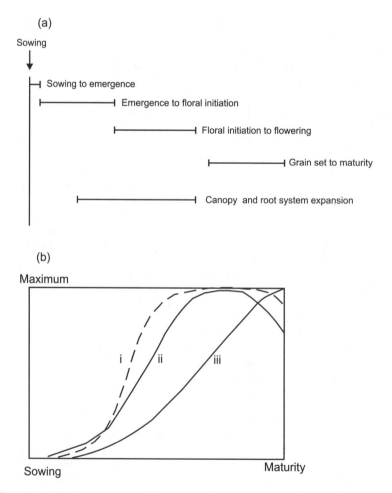

Fig. 5.1. Phenology and resource capture, showing: (a) typical developmental periods in a determinate crop and (b) associated curves for light interception (i), canopy expansion (ii) and dry matter production (iii).

density and solar radiation, while the corresponding rates of leaf expansion or dry matter production per plant vary widely in response to most weather factors and stand composition.

5.1.2 Determinacy and synchrony

A full definition of the developmental habit of a plant must include traits measured at a higher degree of organization than developmental states. Development might progress on only one stem or perhaps in parallel on two or more branches. If the latter, it might occur more or less in phase (synchronously) on the different branches, or else out of phase (asynchronously). On any single branch, development might be closed (determinate), or open (indeterminate), while it might be closed or open on the plant as a whole.

The range of plant forms used in tropical agriculture is very wide. Most modern cereal cultivars are virtually single culmed and entirely determinate; they die when the single reproductive structure is mature. Some landraces of dryland cereals such as sorghum and millet produce many culms, each of which is more or less determinate, but which develop asynchronously. The plant as a whole may be determinate in some genotypes, but partly indeterminate in others, in that new growth will repeatedly spring from the base of the plant (as in many grasses). Legume species, such as groundnut, display a range of habit: mostly asynchronous, with leaves and flowers produced on several branches, but with different degrees of determinacy on different branch-

es. Some of the palms, such as the oil and coconut palms, are single stemmed, but indeterminate, leaves being produced continuously and fruits periodically for many decades. Others, such as the sago palm are again single stemmed but strongly determinate, dying after flowering for the first time. Some common types of developmental habit are represented in Fig. 5.2.

The optimum combination of synchrony and determinacy will vary widely from one cropping system and climate to another. As noted in Chapter 4, plant breeding, when assured of high farm inputs, has moved habit towards high degrees of determinacy and synchrony. Nevertheless, asynchrony and indeterminate habit are strong features of many traditional cereal genotypes, and more generally of a wide range of legumes, root crops and fruit crops. Both synchrony and determinacy are very important in agronomic analysis. The degree of synchrony will often influence the susceptibility of a crop to a stress, while the type of determinacy will affect not just the size but also the direction of the response of a crop plant to temperature gradients, as discussed later in this chapter.

One further point is raised here concerning organization, at the level of the crop population. From germination onwards, there is a range of developmental states among plants within a field. After emergence, the spread of developmental states usually widens, but is sometimes checked by synchronization of flowering in response to a global environmental signal such as photoperiod (Hay and Kirby, 1991). Such spread of development both within and between plants is important to agronomic

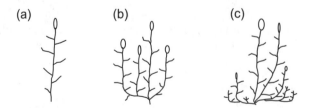

Fig. 5.2. Representations of developmental habit: (a) single stem, determinate, (b) many stems, each stem and whole plant determinate, synchronous, (c) many stems, each determinate, stems asynchronous, plant indeterminate.

practice but has not figured largely in crop physiology and experimental agronomy. An example of convergent development is given later in this chapter, but we return to the matter of variability in development in the final chapter, mainly to emphasize its importance in subsistence and other low-input systems.

5.2 Development and resource capture

The phenology of a crop, together with the climate, sets the potential for resource capture over the season. In determinate plants, the resource capturing surfaces expand mainly in the period up to flowering, before which leaf area index and root length generally reach their maximum dimensions. Very often, senescence during grain filling results in loss of surface towards the end of the cycle. In indeterminate plants, the capturing surfaces can continue to expand after flowering has begun, but once surfaces reach a certain size, further increase suffers a diminishing return (see Chapter 2).

Representative curves of intercepting surface, resource capture and dry matter accumulation are shown in Fig. 5.1b in relation to the developmental periods in Fig. 5.1a. Among most crops, there is a consistent difference in the way these factors change over time in relation to the underlying development. The curve for intercepting surface, e.g. leaf area index, L (curve (i) in Fig. 5.1b), usually peaks just before flowering in most determinate cereals, though in indeterminate plants will later reach some balance between leaf (or root) production and decay. If the crop forms a dense canopy, the fraction of resource it is able to capture on any day (curve (ii)) will rise more steeply and approach its maximum earlier than the curve for surface area. This is because of the diminishing returns caused by self-shading of lower leaves by upper leaves. A canopy typically approaches maximum fractional interception when L is between about 3 and 4, depending on leaf orientation. So if the leaf area index continues up to, say, 6, then the doubling of L will result in only a slightly higher fractional interception. (Similar considerations probably apply to root systems.) Finally, there is usually a fall in curves (i) and (ii) as senescence reduces the area of green leaf. The curve for instantaneous or daily dry matter production (not shown) is similar to that for fractional interception (i), with the exception that shortage of water or nutrients during grain or fruit filling can reduce dry matter production, even if the stand retains a large canopy. The cumulative dry matter of the stand (curve (iii)) lags behind the others because it does not begin to increase at a maximum rate until curve (ii) approaches its maximum.

Manipulation of curves (i), (ii) and (iii) in Fig. 5.1b in relation to the more stable and predictable state-changes in Fig. 5.1a is the basis of agronomic strategy. In this respect, two features of curves (ii) and (iii) have universal significance. The first feature is that, because little growth occurs during the early lag before (ii) starts to rise, the total dry matter production depends on how long curve (ii) remains at or near its maximum. As was discussed in Chapter 4, plant breeding has had most effect on the length of the periods between emergence, floral initiation and flowering (or heading). The effect of such genetic change is to alter the length of time when curve (ii) is near or at the maximum. The resulting genotypic differences in total resource captured and dry matter accrued are among the most substantive of the achievements of plant breeders. A comparison of sorghum varieties in Nigeria (Goldsworthy, 1970) showed that maximum stand dry matter varied from 2.5 t ha^{-1} in a late variety to 1 t ha^{-1} in an early genotype when grown at the same density. Dates of floral initiation were 84 days and 23 days for late and early, respectively, and date of harvest 181 days and 83 days. (We return to this example later in this chapter.) Few authors have measured the effect of such genotypic differences on resource capture, but research in Trinidad on pigeon pea provided direct comparisons of several early and late genotypes (Hughes and Keatinge, 1983). In a series of trials, the cumulative total solar radiation intercepted ranged among early and late genotypes from 300 to 600 MJ m^{-2}

at flowering, and from 650 to 1000 MJ m^{-2} at maturity. Most of the genotypic difference in resource capture had therefore occurred before flowering, as in cereals.

Such large changes in resource capture have important implications for introducing short-season cultivars into a cropping system. Though one of the main advantages of these cultivars is their large potential grain sink, they will only achieve a large biomass if they have a short lag and rapid rise in curve (ii). These can only be achieved at high density plantings and when leaf expansion is not limited by shortage of water and nutrients.

The second important feature of the curves in Fig. 5.1b (whatever the genotype) concerns the balance in resource capture before and after flowering. A more or less equal balance is required to optimize yield in determinate cereals, where the size of the grain sink depends on the rate of resource capture before flowering. The pre- and post-flowering balance is less important in indeterminate plants, whose reproductive development, even after flowering begins, can still respond to temporal change in resource capture. Many of the examples in the rest of this chapter illustrate the advantages of the getting the balance right.

5.3 Environmental control of timing

It is of great advantage to crop management that much of a crop's phenology is controlled by reasonably simple and reproducible responses to two main variables: temperature and photoperiod. Water supply exerts strong effects in some circumstances, especially after sowing and at flowering in some crops, but population density and nutrients are mostly weak determinants of developmental rate in crops (Squire, 1990). Broad responses to the main global environmental signals are now summarized. Case studies are then examined to demonstrate the effects of altitude on the temperature response, and of latitude and season on the photoperiod response.

5.3.1 Global signals

Temperature

Temperature provides ubiquitous control of most developmental processes. Genotypes will develop within a defined range, typically 10–45°C for tropical species. The time for a process typically varies in a U-shape, being very long at both low and high temperatures. The analysis using $1/t$ (Fig. 2.4; Chapter 2) usually linearizes this U-shape so that base, optimum and ceiling temperatures can be identified. If temperature remains at or below the base temperature, development will not occur. At temperatures a few degrees above the base, it occurs sufficiently rapidly to allow commercial production of some crops, such as tea and maize at high altitude. In reality, few crops grow well at or above the optimum temperature, not because of the temperature, but because it is often dry at such high temperatures. Few crops survive if temperature repeatedly approaches or rises above the maximum. If temperature drops below the base during growth, gross development, such as new leaf or flower production, is suspended. For the plant, time effectively stops until the temperature rises again. Thermal time analysis, in which the time–temperature integral is calculated above a base temperature (Chapter 2) is often useful as a means of normalizing a chronological sequence of development or expansion processes.

It is essential to recognize that different structures can be primary temperature-sensors at different stages in the life cycle. In cereals and grasses, for example, the leaf meristem is at first at the base of the plant, initially just under the soil surface, so is influenced by soil and leaf temperature. The meristem later rises, where it becomes closer to air temperature. During grain filling, the sensor is probably at the grain site itself. Knowledge of sensory position is particularly important in analysis of time in cereals and some other plants, whenever there is a marked change in soil or air temperature during the season (see later) or a difference caused by mulching or other treatments to soil.

Photoperiod

Photoperiod affects most (not all) crop species, and usually alters time to flowering, or more specifically, the time to initiation of a reproductive apex. In cereals, flowering itself occurs some weeks after initiation, but this delay is itself less influenced by photoperiod. Plants become sensitive to the daylength after an initial juvenile period. The 'shape' of the sensitivity can then vary widely from one species or genotype to the next.

Photoperiodic effects shorten or lengthen the time that a crop spends near or at the top of the resource capture curve in Fig. 5.1b. In photoperiod-sensitive plants, therefore, change in photoperiod has comparable effects on resource capture to changes in earliness brought about by selection and breeding. The 'short-day' nature of most tropical species means that they either need a minimum short daylength before they will flower at all or that they flower more quickly in short days. Daylength sensitivity is, like temperature sensitivity, reproducible out of context in growth rooms and has been characterized for many legumes and cereals.

Some of the most visible effects of decreasing photoperiod are that plants produce fewer leaves, become shorter in stature and have a smaller leaf area index. Daylength does not affect the rates of these processes directly. Leaf production rates are much the same in short and long days up to the time of floral initiation. The main effect of short days is to stop leaf production earlier. For the most part, temperature affects leaf number and plant height through its influence on the rates of leaf initiation and leaf appearance before an apex or branch turns reproductive. Therefore, most leaves will be produced when temperature is at the optimum, T_o, and fewer are produced as it decreases either side of T_o.

5.3.2 Temperature effects through altitude

Altitude influences plants indirectly, through associated factors such as temperature, rainfall, cloudiness, humidity, windspeed and soil quality. Temperature is the universal factor. The others tend to be locally important to agriculturalists. Along some altitudinal gradients in Africa, for instance, low altitudes are also areas of low rainfall and high saturation deficit. At the other extreme, high windspeed and thin soil might limit crop growth.

The effects of temperature over the usually broad intermediate range of altitude act through the physiological processes described earlier in this chapter and in Chapter 2. While the underlying responses are common to most sites and crops, their actual effect on the way yield changes with altitude is also strongly influenced by other factors, notably the degree of determinacy of the plant and the amount of resource it needs. These factors often interact with each other and with temperature to give complex responses. Agronomic analysis can tease these apart to reveal simpler functions, as is now demonstrated for maize and tea.

Maize in East Africa and Mexico: the determinate condition

In the 1968/69 East African Maize Variety Trials, 23 genotypes supplied by breeders from four countries were trialled at a range of altitudes in six countries in East and Central Africa (Eberhart et al., 1973). Many of the high yielding sites were in Ethiopia and Kenya, while the lowest yielding sites were farther south. There was also a very large range of yield among sites of similar altitude (Fig. 5.3). Nevertheless, regression analysis was used to show that the mean yield of the cultivars apparently increased with altitude by about 700 kg ha^{-1} per 1000 m. Associated with this change was that the duration of the vegetative phase (up to tasselling) became longer with increased altitude, by about 22 days per 1000 m rise in altitude. Crops flowered earlier and matured earlier at lower altitude. Trials in 1970/71 confirmed these responses (Darrah and

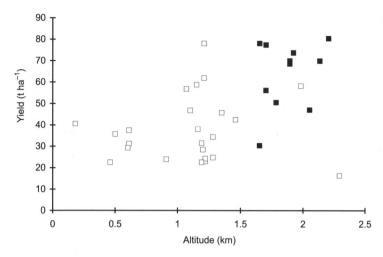

Fig. 5.3. Variation of maize grain yield with altitude and location during the 1968/69 East African maize variety trial: open symbols, Kenya and Ethiopia; closed symbols, Tanzania, Uganda, Malawi and Zambia. (Original data in Eberhart *et al.*, 1973.)

Penny, 1974).

The main causal agent responsible for the trend in these trials was thought to be temperature, which decreases in Eastern Africa by just less than 7°C for a 1000-m rise in altitude. Photoperiod and local environmental factors did not change systematically. The conclusion was that lower temperature delayed development, which in some way allowed an increase in yield (and vice versa). However, the clustering and scatter of coordinates in Fig. 5.3 demands that the effects of altitude should be confirmed inde-

pendently of other environmental factors; and indeed, later, and with benefit of an improved knowledge in crop physiology, Cooper (1979) revealed the physiological links between altitude, temperature and yield.

Cooper (1979) grew one genotype at three altitudes in western Kenya which had a similar photoperiod but differed mainly in temperature. Solar radiation was slightly less at the highest site (but its effects could be accounted for), while soil conditions were generally not limiting. As altitude

Table 5.1. Effects of altitude on phenology, grain yield and vegetative structures in maize grown in Western Kenya; climatic data and locations shown in Fig. 5.4. (After Cooper, 1979.)

Site	Chemelil	Kitale	Elgon
Duration (days):			
50% tassel to maturity	69	83	96
Leafing:			
final leaf number (rounded)	24	23	22
area of leaf 15 (cm^2)	950	1030	1020
leaf area index at 50% tassel	4.6	4.9	4.4
Mean plant growth rate (g day^{-1})	4.8	5.0	4.2
Grain:			
grains per plant	401	481	594
yield per plant (g)	131	166	239

increased (temperature decreased) development became slowed, but the effect this had on different structures was not consistent (Table 5.1). Generally, production and extension of leaves were slowed as altitude increased, but because the duration of vegetative development was also longer, the final leafing traits were not strongly affected by altitude. The duration of grain filling also lengthened with rise in altitude, but in contrast to the response of leaves, the rate of grain filling hardly changed with altitude. Therefore, the final grain weight was mainly proportional to the duration of grain filling and increased with rise in altitude (Fig. 5.4).

The different behaviour of leafing and grain filling in this example probably lies in their different sink requirements. The production of new leaves, and even the extension of leaves, require much less dry matter than the growth of grain. Moreover, the source and sink in the leaf are very tightly coupled, in that a high sink demand (new leaf material) leads very quickly to an increased source (more light captured). For leaves, a longer duration compensated for a slower rate at lower temperature. For grain, however, growth rate was limited by incoming energy; once the canopies had enlarged enough to intercept most of the incoming solar radiation, plant and crop growth rate did not differ systematically between sites, indicating that altitude had little effect on

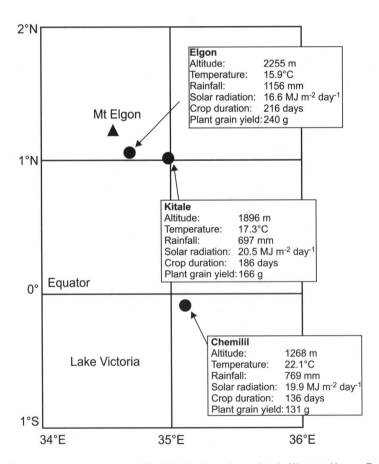

Fig. 5.4. Variation of climate and grain yield with altitude at three sites in Western Kenya. Rainfall was not limiting, radiation differences were small, the main determinant was temperature. See also Table 5.1. (Original data in Cooper, 1979.)

photosynthesis per unit of radiation intercepted. Accordingly, the effect of altitude on duration alone controlled the altitudinal effect on grain yield.

The comprehensive agronomic and physiological research on maize in East Africa leads to the conclusion that yield of a genotype rises as altitude increases because there is more time (and hence more radiation captured) to accumulate dry matter and fill grain. The performance of maize at two altitudes in Mexico (Goldsworthy and Colegrove, 1974) is consistent with this conclusion, but the link between altitude and yield was somewhat different from that in Kenya. In Mexico, unlike in Kenya, mean temperature changed during the crops' lifetime. It was cooler at higher altitude only during vegetative growth, but similar at the two altitudes during grain filling. Even so, yield was greater at the higher altitude. In this instance (as in Kenya), a greater grain sink was set at high altitude in consequence of the lower temperature extending vegetative growth and allowing the stand to accumulate more dry matter at the start of flowering. The durations of grain filling were then very similar at the two altitudes. Grain yield was greater at high altitude because there was more stored dry matter that could be retranslocated to the sink.

One way or another, the response of grain yield to altitude at both of the sites was underpinned by the very simple response of developmental time to temperature. Again at both sites, this response led to more grains at higher altitude (cooler temperature). The physiological link between developmental time and grain number, through thermal growth rate, is explored elsewhere (Cooper, 1979; Hawkins and Cooper, 1981; Squire, 1990). The question to consider now is whether yield will always increase with rise in altitude, other factors being non-limiting. In both the studies referred to here, the relation held because the source, including stored dry matter, and the grain sink were in balance. The relation is most likely to break down when they are not in balance. Examples of this include: (i) when temperature is so low at high altitude that it damages fertilization and grain set, but has little

effect on vegetative growth and photosynthesis; (ii) when pests, disease, competition or senescence reduce the intercepted radiation or photosynthesis rate after sink-set *and* when the genotype can retranslocate little of its stored assimilate. Such effects and mechanisms will interfere with the general response locally but will not overturn it. If we look beyond cereals such as maize, however, *this rise in yield with rise in altitude* is not universal, and is the norm in cereals only because of their highly determinate habit. A contrasting response is now shown by reference to an indeterminate perennial plant.

Tea in East Africa: the indeterminate condition

It has long been recognized that tea yields decline with rise in altitude, a response opposite to that just described for maize. The yielding structure in tea is a small shoot, usually 15–20 cm long, supporting two or three leaves and a terminal bud. When this shoot is plucked, one or sometimes more, buds below it are released from apical dominance and themselves begin to grow. Yield per unit field area therefore depends on the size of the basal shoot population, the time it takes for each shoot to grow from a bud to harvestable size and the weight of the shoot at harvest (Squire and Callander, 1981; Carr *et al.*, 1987). In various parts of the world, tea experiences a range of environmental factors related to altitude, including thin soils on steep upland slopes and humidity-related disease in low lying areas. Over some parts of the tea region in East Africa, however, the soils and other climatic factors are not strongly limiting and the main altitudinal effects operate through temperature.

The effects of temperature on tea shoot growth as a sink-limited process were first systematically examined in Malawi in relation to seasonal change in temperature (Squire, 1979; Tanton, 1982a, b), but effects of temperature were difficult to separate completely from those of rainfall and humidity. Later, measurements at several altitudes in Kenya revealed that the time for a shoot to reach harvestable size was much

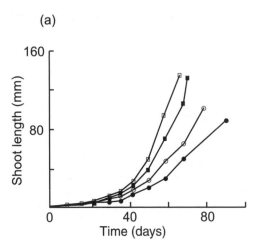
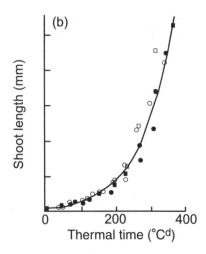

Fig. 5.5. Growth in shoot length of the tea clone S15/10 at four altitudes in Kenya, expressed in (a) time and (b) thermal time above T_b = 7.5°C. The curve in (b) is a fitted exponential. Symbols: (□), 1860 m; (■), 1940 m; (○), 2120 m; (●), 2180 m. (Adapted from Squire *et al.*, 1993.)

more strongly influenced by temperature than were the other two components of yield, the shoot population and shoot mass. The time from bud break to harvestable size in tea was shown to be related to temperature by the same type of response as leaf production and other development events in maize and other annuals. For most of the year, the growth of shoots at different altitudes could be normalized by expressing extension growth in cumulative thermal time above a base temperature of 7.5°C (Fig. 5.5). Each shoot cycle, from bud break to the harvestable size, took almost 90 days at 2200 m compared with 65 days just 300 m lower. Over a year, therefore, there were four shoot cycles and four shoots plucked at the higher altitude compared to almost six cycles and six shoots plucked at the lower.

Tea shoots can now be compared with both maize leaves and maize grain. They share the property with maize leaves that a shorter duration of growth at low altitude is compensated for by a faster rate of extension because extension rate of the leaves is seldom limited by the source of assimilate. Accordingly, the final length of a shoot (and so its mass) is not much affected by altitude and temperature. Tea shoots differ from maize leaves in having indeterminate cycli-

cal growth: while only one leaf is produced at each node in maize irrespective of temperature, more shoots are generated per year at each position on the tea bush surface as altitude falls (temperature rises). Tea shoots differ from grain growth in maize by being not limited by the source of assimilate.

A general point to note when comparing genotypic differences

In these examples, a simple response to temperature underlies the main trends of yield with altitude, even though the size and the direction of the trend depended on the species' developmental characteristics. It is now possible to make a general statement about the way *a difference* between two genotypes or two cultivars of similar habit changes with altitude. This has practical importance because in many multi-site genotype by environment trials, the difference in development and yield between genotypes rarely remains constant, but diverges with change in altitude, and this divergence itself is taken to have some genotypic significance. In the East Africa maize trials, for instance, cultivars were more similar in the time to flowering (and thereby in yield) at sea level than at 2000 m. However,

Table 5.2. Calculated effect of temperature on the time between the start of canopy expansion and flowering in two hypothetical cultivars

Altitude (m)	Temperature (°C)	Cultivar A θ = 700°C	Cultivar B θ = 500°C	Difference (A − B)
Low	25	47	33	14
Medium	20	70	50	20
High	15	140	100	40

this divergence is simply a consequence of the shape of the response of development to temperature.

The point is illustrated by two hypothetical cultivars which have a similar phenology but different thermal requirements. They all have the same base temperature (T_b = 10°C) but a different thermal duration for the developmental period between the start of canopy expansion at flowering (θ = 500°C^d, 700°C^d). The time required to pass through this developmental period ($θ/(T−T_b)$) is shown at three temperatures in Table 5.2. The time increases with rise in altitude (fall in temperature) for both cultivars but the difference in time between them also increases. This increase in the *difference* occurs simply because the developmental time is inversely proportional to temperature above a base, not because the process in one or other cultivar is in any

other way more sensitive to temperature.

5.3.3 Photoperiod effects through latitude and season

The example of maize in East Africa worked as a demonstration of altitudinal effects because the associated temperature gradient was the only systematic environmental factor influencing yield. Comparable demonstrations of the effects of latitude on phenology are much more difficult to find because other factors, such as rainfall, vary with latitude in many parts of the tropics. The results from a series of trials and experiments with sorghum and maize in West Africa are now drawn on, since they exemplify effects that can be seen or inferred elsewhere but have rarely been so well documented (Fig. 5.6). The ecology of the region

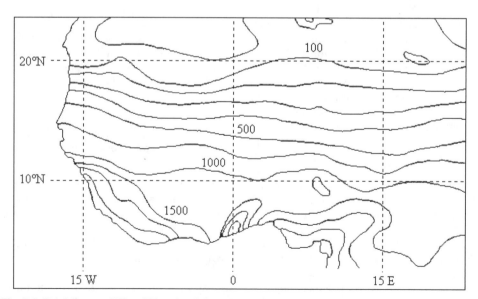

Fig. 5.6. Rainfall map of West Africa showing contours of mean annual precipitation labelled at 100, 500, 1000 and 1500 mm.

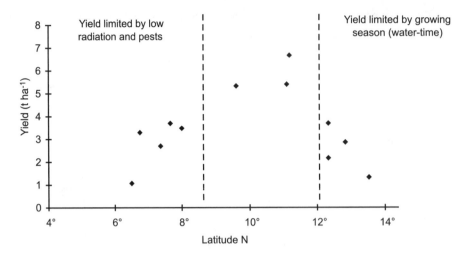

Fig. 5.7. Variation in maize yield in West Africa with latitude from the coast to dry savanna in the north. (Adapted from Kassam *et al.*, 1975.)

is described by Kowal and Knabe (1972).

Extensive information on yield and latitude was obtained from multi-location trials in the 1960s and 1970s in various countries including Dahomey, Ghana, Nigeria, Ivory Coast, Sierra Leone, Burkina Faso, Mali, Senegal, Gambia and Niger. The yield of maize averaged for several cultivars increased with latitude from 6° N to a peak between 10° N and 12° N, then declined with further increase in latitude to 14° N (Fig. 5.7). Kassam *et al.* (1975) explained that the maximum mean yields were obtained in the Guinea savannah region (between 8 and 12° N) where the rains were long enough to allow most genotypes to reach maturity. To the north of this (the Sudan savannah), the incoming solar radiation increased with latitude, but the shortness of the rains limited growth and thereby led to wastage of radiation. To the south (the forest and derived savannah), the climate was wet enough for water not to limit growth, but the associated cloudiness (low radiation) and increased pest problems did limit it. Throughout this region, local cereal genotypes had become adapted to the latitudinal change in conditions so that they matured at about the end of the rains. New cultivars, whose phenology was not adapted, yielded less than the local ones whenever they matured substan-

tively earlier, because of either pest problems, or later because of lack of water (Curtis, 1968).

A further phenomenon was that a variety reached maturity within a short period within the annual calendar to a large extent irrespective of the time of sowing after the beginning of the rains. This convergence to a point in time was considered an adaptive response to photoperiod because the date the rains began was more variable from year to year than the date the rains ended. Matching phenology with the end of the rains was more important, therefore, than matching it with its beginning (Andrews, 1973).

Convergence in maturity caused by decreasing photoperiod

The above observations prompted an experiment at Samaru (11° 11′ N) in which the photoperiod-sensitive sorghum variety, Short Kaura, was sown at ten different times in May, June and July (Kassam and Andrews, 1975). The temperature did not change much during the season, but the photoperiod varied from a maximum of almost 13 h on 21 June to a minimum of 11.5 h near the end of November (Fig. 5.8a). The rains began in April and finished at the beginning of October. The first point to note is that the range of dates for floral initiation, heading and maturity became increasingly compressed, to a degree that the range for maturity was less than a quarter of that for sowing. The total life cycle decreased from 188 days (first sowing) to 123 days (last), much of which change was caused by a decrease in the vegetative phase (i.e. earlier floral induction) from 98 to 59 days. The second point is that grain yield decreased from a maximum of 3–3.5 t ha^{-1} in the first few sowings to a very low yield in the last sowing.

The interpretation of the results was that later sowings developed in shorter photoperiods which acted to decrease the time after emergence when the vegetative

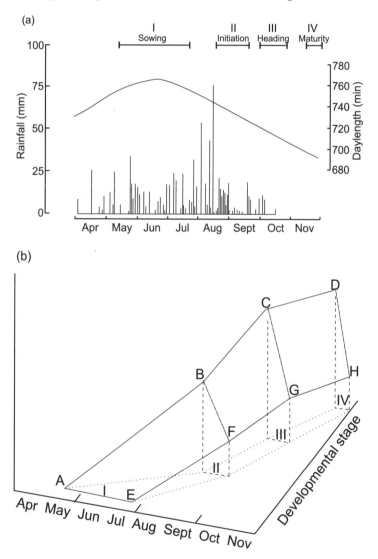

Fig. 5.8. Convergence of development for sorghum sown at 10-weekly intervals at Samaru, Nigeria: (a) seasonal rainfall, daylength, and ranges of sowing date and developmental periods (original data in Kassam and Andrews, 1975); (b) three-dimensional representation of effect of developmental convergence on plant size and mass (vertical axis) showing early sowing (ABCD) and late sowing (EFGH) and time differences at sowing (I), floral initiation (II), heading (III) and maturity (IV).

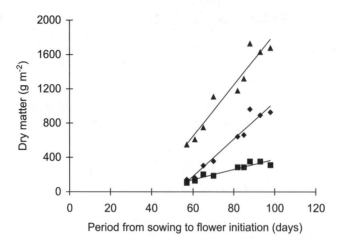

Fig. 5.9. Effect of developmental convergence on sorghum biomass in the experiment in Fig. 5.8: standing biomass at floral initiation (♦) and heading (▲); grain yield at maturity (■). (Original data in Kassam and Andrews, 1975.)

apex turned reproductive. That this did not happen on exactly the same date implied that the photoperiodic signal was not a sudden response occurring after a particular short daylength has been passed, but a cumulative effect with the shorter days having more 'weight' as the season progressed. The earlier floral induction in later sowings prevented full expansion of the canopy and thereby reduced dry matter production (Fig. 5.8b and Fig. 5.9). With reference to Fig. 5.1b, the curve (ii) for later sowings did not reach its potential maximum and the whole cycle, moreover, was cut short. As in the responses to altitude described earlier, grain number and grain yield were strongly dependent on the rate of resource capture and dry matter production in the weeks leading up to flowering. Given the much smaller canopy size and resource interception of the later sowings, they were quite unable to set and fill a large sink.

5.4 Regulation of timing by management and choice of cultivar

As a first approximation, temperature controls development much as a clock, ticking towards a predetermined time in the future. As the clock ticks, new cells are produced and extend, differentiate and fill. In response to photoperiodic signals, the developing meristems might become reproductive rather than vegetative, but still move on according to the temperature-determined clock. To an extent that differs among genotypes, the clock keeps to time, even though processes of expansion and dry matter production might be suppressed by shortage of water and nutrients.

5.4.1 Matching crop phenology with temporal change in resource

One of the main aims of manipulating timing in crop production is to match this stable phenological timing mechanism with the expected temporal change in the available physical resources. The solar radiation and rainfall, the nutrient released from soil, the daylength and temperature and whether a field can be cultivated, are all coupled to a degree, but rarely are all maximal at the same time. The timing of sowing and harvest, and of any intervention to modify the main developmental transitions, are therefore compromises, designed in low input agriculture to waste as little as possible of the limiting resources.

Types of resource loss

It is feasible for most crops to construct a balance sheet of potential losses. It can then be considered where most gain (least loss) is likely to be made given available genotypes and options for management. The curve in Fig. 5.1b, showing change of fractional resource capture over time, can be re-drawn to show how loss of resource can be considered during each of four periods (Fig. 5.10). In the first instance, we can assume that the resource is solar radiation. The first period of loss occurs while the leaf surface is small and expanding, and is inevitable in stands of single species annuals ((I) in Fig. 5.10). The shape of the rising curve is such that interception during this first phase is typically 35% and losses 65% of incoming radiation (Squire, 1990). The second period occurs when the canopy has reached maximum size. Loss here (II) results from solar radiation passing through the canopy to the soil and depends on the final size and the dimensions and architecture of the canopy. The third period is after senescence begins to reduce the area of surfaces (III). The losses at maximum intercepting area and during senescence can be virtually eliminated given no limit to inputs. The fourth period of loss (IV) occurs after the crop is harvested, and will cover the time when resources are still available up to the end of the potential grow-ing period as defined by the onset of drought or frost. Hughes (1986) quantifies the magnitude of I–IV for a range of chickpea crops.

A similar scheme can be applied when water is the main limiting resource. Early losses (I) occur as evaporation directly from the soil surface before the root system has fully expanded. Loss of type II approximates to the water remaining throughout the soil profile when the root system has taken as much water as it can when at its maximum dimensions and root length density. Losses during senescence (III) and after maturity (IV) occur potentially as for the shoot system. The balance sheet of losses for radiation and water are different, however, in that radiation not intercepted by a canopy can never be regained, while water inaccessible to a root system at one stage might still be absorbed later following further growth of fine roots. Many dryland crops are potentially able to eliminate losses analogous to type I in Fig. 5.10, except those by soil surface evaporation. The management options to manipulate water losses are discussed in Chapter 7.

In dryland cropping, management has the scope to reduce loss of water in several ways. For instance, soil surface evaporation can be reduced by mulching or land shaping techniques (see Chapter 7) while tillage to reduce soil bulk density and to break up hard layers increases the amount of avail-

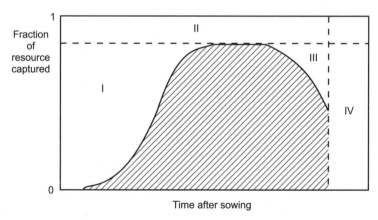

Fig. 5.10. Representation of seasonal change in fraction of resource captured showing losses during canopy expansion (I), at maximum expansion (II), as a result of senescence (III), and after the crop is harvested (IV).

able water, and so reduces type II loss. More strategic thinking is required however in order to deploy a combination of cultivar and stand density that minimizes type IV loss by depleting the resource as near as possible to maturity.

Cultivar phenology and water availability

The period of the year when crops can be grown varies widely in the dry tropics, depending on the onset and length of the rainy season and distribution of rainfall events. Many cropping systems have to sustain a stand over a period without rain, when water stored at depth in the soil is used. This might be in the middle of the growing season where the rainfall distribution is bimodal or at the end of the season if the period of rain is short and the distribution unimodal. In some systems, a soil profile is purposefully flooded by diversion of a river in spate, and the whole crop is then grown on stored water (Chapter 7). Water stored at depth and abstracted at the end of a stand's life is advantageous, as it allows a crop to keep transpiring while its grain matures unwetted. However, the principles linking crop development to water supply are the same whatever the rainfall distribution and the reliance on stored water.

Imagine that there is enough water in a hypothetical environment to allow a stand to transpire for 100 days; this can be considered its 'water-time' (Monteith, 1986b). The ideal developmental time from sowing to maturity of a crop is also 100 days. If a crop reaches maturity in less than 100 days, it wastes water. If it reaches maturity well after 100 days, it might die or fail to yield. Management has two broad options to achieve the best match between water-time and developmental time. First, as Chapter 4 concluded, plant breeding has provided genotypes with a wide range of phenology, albeit influenced within limits by effects of temperature and photoperiod. Second, choice of stand density is a very effective way of regulating water-time.

To demonstrate the importance of genotype, we return now to data from experiments along the ecological gradient in West

Africa. Trials with sorghum in northern Nigeria between 1957 and 1961 showed how crucial is this match of developmental time and water-time. Farmers in each ecologically distinct area had developed their own sorghum varieties which were reputed to grow less well or to fail outside their own area. To examine the basis of this effect, sev-

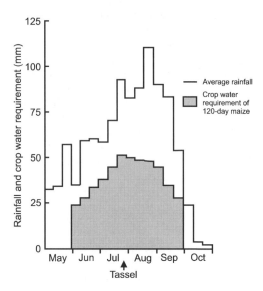

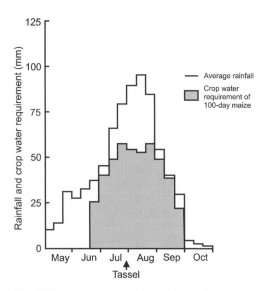

Fig. 5.11. Average rainfall at two sites in West Africa and measured water requirement of maize cultivars whose phenology fits the season at each site. (Original data in Kassam *et al.*, 1975.)

eral 'standard' varieties, along with the local cultivars, were grown at a range of sites along the (latitudinal) dryness gradient. The standard variety 'Yar Gunki', for example, yielded similarly to the local cultivars only when it headed at about the same time as them. In the northern, dry areas (Sudan Savanna) it developed later than the local varieties and in all but one trial, either flowered and yielded little, flowered but yielded nothing or did not flower (Curtis, 1968).

When species, such as maize, that have not been habitually grown in this region, are introduced or encouraged, then clearly one genotype will be inadequate to cover the whole range of conditions. This is shown by a comparison of rainfall and measured transpiration of crops grown in lysimeters at two points along the latitudinal gradient in Figs. 5.6 and 5.7 (Fig. 5.11). At about 11° N, the rainfall distribution allows a 120-day maize to be grown between the beginning of June and the end of September. At 13°N, the 120-day maize would fail, but a 100-day genotype neatly fits the reduced growing season (Kassam *et al.*, 1975).

Microclimate, stand density and water-time

Though the microclimate in and around a stand has a strong influence on both developmental time and water-time, there is generally little scope for management to modify it, at least in single storey stands. The effects of mulch on soil surface temperature are discussed briefly later, but once the effective meristems are in the aerial parts of the plant, phenology is largely under the control of global signals. The transpiration rate is clearly driven in humid conditions by microclimatic factors such as saturation deficit and windspeed, but in water-limited conditions become more dependent on rooting factors.

The principal means by which management can alter water-time is through stand density. Under conditions when the soil surface is dry and the stand is growing on stored water, density controls transpiration through the rate at which a root system descends into soil, proliferates and abstracts water. Generally, the denser the stand, the faster the descent of the root system and the growth of root length density. The most appropriate density for a combination of crop, soil and climate has often been found by local experience or empirical experimentation. However, given knowledge of the typical rooting profile and the volume of water stored in the available range, it is quite feasible to estimate the water-time for a particular stand. (Chapter 6 examines density effects in more detail.) The resulting optimum is usually at least ten times lower than the stand density giving maximum resource capture when water is not limiting. Typically, such a stand will expand its root system at a rate which keeps transpiration at 2–3 mm day^{-1}, somewhat independent of the potential evaporation rate. This rate of water loss is high enough to provide a level of cooling and hydration that prevents foliage from scorching, yet is low enough to extend water-time by two or more weeks for an arable crop growing on stored water.

Optimizing source and sink when water is limited

Given a genotype with a suitable developmental time, and a planting density that provides an acceptable water-time, the next point to consider is how the use of resource has to be balanced among the various developmental phases. At one extreme, we might consider a hypothetical crop that always allocates the same fraction of biomass to yield. It achieves a stable yield whether it uses the available water early or late in the season. In most instances, allocation patterns are far less stable than this, to an extent that balancing the amount of water used in different developmental periods is an important objective. Again, stand density is the universal factor that management can act on. Just how fine the temporal balance of transpiration has to be will be influenced by two main physiological features: the developmental habit, particularly the determinacy; and the extent to which stored dry matter can be retranslocated.

Generally, the temporal balance of water use is more critical for determinate cereals than for most other crops. Evidence was discussed earlier which suggests the rate of dry matter accumulation before flowering strongly determines the number of grains set on a plant (Cooper, 1979; and see Chapter 4). Fischer (1979, 1981) has examined this for Mediterranean crops but his conclusions apply also to tropical dryland cereals. It is proposed that there is an optimum dry matter at flowering, which results from optimum rates of resource capture in the few weeks before flowering. If the dry matter of a stand is below this optimum, it will not set enough grain, and probably not all water will be used by maturity. If above this optimum, it will set more grain than it can fill, primarily because the stand will have used too much of the water before flowering and will suffer drought during grain filling. In extreme instances, it may not be able to fill any grain of adequate quality. Achieving this optimum is one of the most difficult tasks in dryland farming, and we return to it in Chapter 7.

Agronomy can be much more relaxed about the temporal use of resource, when a genotype either has indeterminate reproductive development or can retranslocate stored assimilate to fill grain. Where legumes and storage crops use only current assimilate to fill pods, there is not such a need to balance pre- and post-flowering resource use. Fruit trees, such as the oil and coconut palms, have intermittent development where the fruit sink is matched with the current source through internal control of the gender of the inflorescence and by abortion of young fruits (see later in this chapter). The capacity of plants to retranslocate is a more uncertain phenomenon in that it does not seem to happen consistently. Ludlow and Muchow (1990) rated retranslocation as the third most important trait to introduce to dryland cereals. If the capacity were absolute, say half the stem mass could be retranslocated, then the strategy of management would be to extract the water for just long enough for grain to be set. Yield would then be independent of the finer details of transpiration rate during the season. In reality, there are many instances when crops have accrued massive amounts of dry matter in their stems but have moved none of it to the panicle. Goldsworthy (1970) gives examples of this in long-season traditional varieties of sorghum, but there is elsewhere ample evidence that retranslocation occurs in sorghum. Perhaps the uncertainty lies less in the capacity for translocation, which is probably always there in maize, sorghum and millet, than in the conditions that lead to a large grain sink.

5.4.2 Time of sowing (planting) effects

Some of the clearest examples of time of sowing effects were those where sorghum was grown at latitudes where daylength changed with time of year enough to cause large changes in phenology (Andrews, 1973; Kassam and Andrews, 1975). Delaying sowing not only caused some crops to run out of water, but – through the effect of decreasing photoperiod – caused plants to flower much earlier and before they had accumulated a good canopy and adequate dry matter to fill grain (Fig. 5.8). Where timely land preparation and sowing are possible, natural and human selection have tended to encourage the phenology of crops to fit seasonal windows. Since the developmental clock for a genotype is set at sowing, the time of sowing determines the time of harvest.

In addition to such broad-scale effects, the immediate states of several environmental variables around sowing are also critical. For those sown at the beginning of a wet season, the soil temperature, soil moisture, hardness of surface and oxygen concentration will all be uncertain. Sowing too early might mean loss of crop if seeds germinate then die from drought or high temperature.

Time of sowing effect in the Kenyan highlands

Cooper and Law (1978) report one of the few studies that have tried to trace the physiological origin of time of sowing effects. In the Kenyan highlands, every week's delay in sowing maize after the start of the rains reduces yield by 0.6 t ha^{-1}. These authors established that the grain yield was very

Table 5.3. Environmental and physiological differences between crops sown weekly (1, 2, 3, 4) following the start of the rains at Kitale, Kenya, 1993.

	1	2	3	4
Up to 5 weeks post-emergence:				
Mean air temperature (°C)	18.0	17.7	17.3	16.9
Mean soil temperature at 7.5 cm (°C)	22.2	21.9	21.6	21.3
Solar radiation (MJ m^{-2} day^{-1})	19.4	19.2	18.6	17.9
Plant dry matter (g)	19.9	15.7	14.1	13.1
At tasselling:				
Plant dry matter (g)	286	263	205	169
Leaf area index	4.8	4.9	3.8	3.4
Time, emergence to tasselling (days)	91	91	93	96
Plant grain yield (g)	205	185	159	105

strongly correlated with plant dry matter only a few weeks after emergence. The causal links between early expansion and final grain yield were not fully established but possibly the most likely explanation starts with the observation that soil and air temperature, and also solar radiation, decrease gradually from the start of the rains (Table 5.3). Early sowings therefore develop faster, and achieve a greater canopy surface (and intercept more radiation) leading up to flowering. This enables them to set a greater grain sink, resulting in about twice the grain yield of later sowings. The time from emergence to tasselling was increased by only 5% between crops sown between the first and fourth week after the rains. The time of sowing effect was not mediated primarily through long-term crop phenology but through differential canopy expansion. One unresolved question in this work is how such a small, though highly systematic, difference in mean temperature, caused such large differences in canopy size. One explanation might be that the meristem was near the soil surface, the temperature of which was much higher during the day, early in the season, than either air temperature or soil temperature at a depth of 7.5 cm.

Contrast between early season and whole season temperature effects

In the above time-of-sowing effect, the postulated temperature response is the same type of physiological response that caused Kenyan maize yield to vary with altitude. In both instances, higher temperature caused faster development of leaf primordia and faster expansion of leaf blades. Increasing temperature had opposite effects in the two instances, however. In the altitude trials, higher temperature reduced yield; in the time-of sowing trial, it increased yield.

The contrast between the altitudinal and time-of-sowing responses demonstrates the need to be cautious in interpreting field data when temperature is changing through the season or effects one part of the life cycle much more than another. In the maize altitude trial in Kenya, vegetative and grain filling periods experienced the same temperature. When this is so, all developmental stages decrease as temperature rises (as Table 5.1 shows, and assuming water supply is not limiting, and that both processes have the same base temperature). This means that earlier canopy closure at high temperature never compensates for the shorter duration of grain filling. In the maize sowing date trial, temperature differentials occurred mainly during early canopy development, while maximum stand dry matter, grain filling and maturity occurred at a similar time after sowing. In any situation of this type therefore, higher temperature causing only earlier canopy closure will generally increase yield.

5.5 Conclusions

The emphasis in this chapter has been on the regulation of phenology by global environment. In lucrative and high input agriculture and horticulture, control of temperature and photoperiod is essential and is achieved by enclosing stands in glass or polythene and imposing a daylength using artificial lighting or screens. In most of tropical agriculture, management can do little to modify these factors, so operates to channel their effects by choosing genotypes and by regulating the temporal use of resources through stand density and configuration. We have concentrated on case studies in maize and sorghum but have tried to draw out generalities. A useful student exercise would be to repeat the analysis for grain legumes, root crops or fruit trees. In doing so, there are several aspects of phenology that might be borne in mind, even if they are not yet well understood.

5.5.1 Earliness and photoperiod vs. altitude

Choices of genotype and management are never independent, but certain generalizations can be made about the need to adapt management factors over a climatic gradient. There are probably no universal rules, but over limited parts of the tropics there would appear less need to change stand density over altitudinal (temperature) gradients than latitudinal (photoperiod, dryness) gradients. The point to determine is whether vegetative characteristics such as plant height and leaf area will change or not. For a determinate cereal grown over a range of altitudes, the evidence presented here shows they will not change by much. This may well mean that the optimal density will also not change with altitude and temperature.

The situation is therefore quite different for the response to photoperiod, where one of the main effects of short days on a short-day tropical crop species is to reduce plant height and leaf area at flowering. Introducing an early cultivar to replace a late

one has the same implications. In both instances, optimal density is likely to be greater for the vegetatively less vigorous early cultivar and the short-day crop growing under short daylength. This is an example of where cultivar and management are not transferable.

5.5.2 Exploiting plants' own stabilizing processes

In adverse conditions, agronomy benefits from understanding and using the plants' propensity for surviving and reproducing. An excellent example is afforded by the sequential sowings of sorghum depicted in Fig. 5.8 and Fig. 5.9. The shortening of vegetative development in response to decreasing photoperiod allowed most stands to reach maturity before all the stored water had been used. The plants in later sowings made (in effect) further corrections that culminated in only a twofold difference between sowings in grain number per plant.

In cereals, the developmental processes leading to grain number are contracted in time; the period from inflorescence initiation to pollination is often similar to that between pollination and maturity. The variety of stabilizing processes is much clearer in a crop such as oil palm, where the period of fruit filling, lasting 1 or 2 months, is a small fraction of the 2 or 3 year period of reproductive development (Fig. 5.12). Summarizing earlier research in West Africa and Malaysia, Corley (1977) demonstrated that the source and sink were balanced by a series of coupled processes that began at least 2 years before the fruits appeared. A low source, caused by drought or heavy pruning for instance, increased the number of male inflorescences and reduced spikelet number. When meristematic activity ceased about 8 months before pollination, further regulation occurred by abortion of inflorescences. Even after pollination, immediate imbalances could be regulated by tolerances in the mass of fruits or by use of assimilate from store in the trunk.

At least some of such processes operate in all tropical crop species. The lesson is

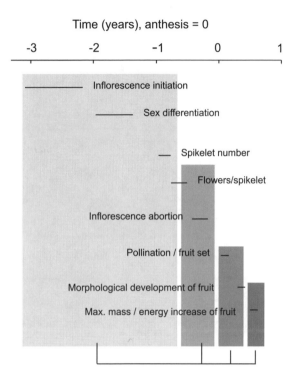

Time (years), anthesis = 0

Inflorescence initiation

Sex differentiation

Spikelet number

Flowers/spikelet

Inflorescence abortion

Pollination / fruit set

Morphological development of fruit

Max. mass / energy increase of fruit

Fig. 5.12. Interactions among phases of reproductive differentiation and fruit bunch growth in oil palm. Time 0 is anthesis. (Drawn from information in Corley, 1977.)

that any agronomic intervention early in the life of a crop will affect later processes to at least some degree. Knowledge of the strength of the coupling between early and late events is one of the prerequisites of working with any species or genotype.

5.5.3 Within-cultivar variability in phenology

We return here briefly to the argument about variability in timing and other plant traits. Natural and artificial selection have tended to remove weedy characteristics such as seed dormancy from crop species and to reduce the spread of time to emergence in a seed lot. However, variability in emergence time can be advantageous when there is an uncertain start to a rainy season. For instance, early emergers, stimulated by a light shower,

might die before the soil below is wetted. A larger seed size provides some security because the greater reserves allow seeds to be sown deeper in the soil where wetting is a consequence of heavier rainfall and where they are subjected to less extreme fluctuations in temperature. The evolutionary trend in crop species towards larger seeds has probably occurred in response to the need for less haphazard emergence as well as ease of harvesting and food preference.

There has, however, been little systematic study of the implications for final yield of variable emergence in relation to early rainfall. Most physiological studies still focus on the mean or mode of the sample and do not follow through individuals emerging at different times to see how they fare later in the season. Within-cultivar variability is considered in the final chapter.

6

Configuration (How?)

There are basically two strategies for matching the availability of *environmental* resources to the *biological* demands of crops. Either more resources can be added to the system or the choice of crop and cropping system must be manipulated to match the resources available. In many temperate environments, the supply of natural resources (e.g. rainfall and nutrients) can be supplemented to meet the demands of a particular crop, for example by the application of irrigation or fertilizers. Although the irradiance above a crop cannot be changed, the fraction and duration of the radiation intercepted by foliage can be maximized by applying fertilizers to increase the expansion rate and duration of the canopy and pesticides and herbicides to maintain healthy leaves and deter competitor species. In these circumstances, there are many options available to the grower in terms of the species to be grown, the population and arrangement of plants and the timing and amount of inputs. In practice, the demands of the crop can be met by ensuring that no factor (other than intensity of radiation) is limiting the biological potential of the crop at any stage of its development.

Ideal conditions for crop growth rarely occur in the tropics because there are limited natural resources and farmers are often too poor to supply crops with additional inputs. The only practical option in many tropical environments is therefore to adjust the biological demands of the crop to match the availability of natural resources expect-ed during the season. An example of such a strategy is the selection of progressively more drought-tolerant species with declining likelihood of rainfall. For example, although farmers in many dry regions might prefer to grow cereals such as maize, in practice they may have to grow species such as sorghum or millet. The greater drought tolerance of these latter crops means that they are more likely to achieve at least some yield where the maize crop will fail altogether.

In practice, the crop manager has two broad options. The first is to adjust the *planting density* of a crop at any site in relation to the weather anticipated in any particular season. In this way the demands of the crop are spread so that the most limiting resource, often water, remains accessible until the end of the season allowing the crop to produce at least some reproductive yield. In contrast, a crop sown at a greater planting density may already have used all the available resources through its greater demands earlier in the season.

The second is to grow two or more species together in mixed cultivation. In these circumstances the intensity and duration of competition for resources depend on the growth habit of each species and interactions between different species growing in close proximity. There are many forms of mixed cropping ranging from *blends*, where more than one variety of the same species are grown together, through to *intercrops*, where two or more species are intermingled, and *agroforestry*, where annual arable crops are grown in association with perennial trees. Although these various combinations of species exhibit a vast range of forms and types, there are certain principles that underlie the source of their productivity. An understanding of these principles may help to explain why mixed crops often use available resources to produce a greater combined yield than their constituent species grown alone.

Both planting density and intercropping are examples of how the *configuration* of crop stands can be adjusted to optimize yield under differing circumstances of soil and climate. After choosing which species to grow and when to sow seeds, identifying the best configuration of a crop stand is often the next priority for farmers in the tropics. This chapter considers how different planting densities and mixed systems capture and convert environmental resources into dry matter and yield. Examples of how the performance of different planting densities or combinations of species can be measured and predicted are also discussed.

6.1 Plant population

The optimum planting density for any crop is usually determined by trial and error. As a result, recommended sowing rates for different crops are based on evidence accumulated from field trials which have been repeated over a number of seasons to account for annual variations in weather. The optimum planting density at one site may not apply at other locations because regional variations in weather and soil type mean that further trials are needed at each site to validate general recommendations. This process is time consuming and labour intensive and the results from such population/yield experiments provide little or no increase in our understanding of *how* crops at different populations capture and use resources to produce a final yield.

An alternative to the approach described above is to evaluate crop productivity in relation to the availability of environmental resources expected during a growing season. For many tropical crops,

the most limiting resource during the growing season is likely to be the supply of water to plant roots. So, it is appropriate for us to review plant population/yield relations in terms of differences in the amount of soil moisture available to crops ranging from circumstances where water is never limiting throughout growth to the extreme case where soil water is always scarce.

6.1.1 Water non-limiting

Most fundamental studies of crop productivity and planting density have been restricted to temperate crops, usually well supplied with water. In these circumstances, the most limiting resource is usually solar radiation. Therefore, with increasing plant population, total dry weight, W, per unit area of ground often approaches an asymptote at a population which almost completely intercepts radiation (Holliday, 1960; Willey and Heath, 1969). This so-called 'reciprocal relation' can be described as

$$\frac{1}{W} = \frac{1}{w_m N} + \frac{1}{W_m} \tag{6.1}$$

where N is plant population, w_m is the maximum productivity per plant when grown under non-limiting conditions and W_m is the maximum productivity of the whole crop.

The value of W_m for a particular crop growing at any site without nutrient or moisture limitations depends on seasonal radiation. This situation is schematically represented in Fig. 6.1. Here, the total dry matter of a crop is usually well correlated with seasonal accumulated intercepted radiation (see Chapter 2). Because radiation is the driving force for evaporation, the asymptote for total dry matter against population is also similar to the seasonal value of the reference evaporation (E_{to}). This link holds because, except for the period before complete canopy closure when losses directly from the soil are significant, the bulk of total evaporation can be accounted for by transpiration and in these circumstances, crop

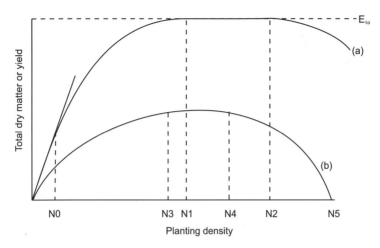

Fig. 6.1. The idealized relation between total dry weight (a) and yield (b) of a crop with increasing planting density when water is not limiting. N0, planting density at which competition between individual plants begins under non-limiting soil moisture; N1, planting density at which crop dry weight reaches a plateau; N2, planting density at which dry weight begins to decline; N3, planting density at which yield reaches a plateau; N4, planting density at which yield begins to decline; N5, planting density at which yield declines to zero; E_{to}, reference evaporation.

transpiration, reference evaporation (E_{to}) and potential evaporation (E_o) are all similar (see Chapter 3).

Total dry weight

If we consider the shape of the relation between total dry weight and increasing planting density (Fig. 6.1) it can be seen that there is initially a linear slope where each additional plant contributes its maximum potential productivity (i.e. w_m) set by its genetic composition with no apparent competition from its neighbours for resources. The linear relation between dry matter and increasing population begins to decline (N0) when individual plants start shading each other and compete for light. Thereafter, there is a diminishing return of dry matter for each extra unit of intercepted radiation. Once complete ground cover is achieved, any potential increment in total dry weight through an increase in the number of individual plants is exactly matched by the decline in dry weight per plant, because plants are forced to share a finite amount of resource. Therefore, beyond population N1 there is no advantage in increasing planting density, since no greater total biomass can

be achieved for a particular crop at that location. In fact, there may be a real disadvantage in any substantial increase in plant population. Even when the total soil water is not limiting, at very high planting densities (N2) intense local competition for resources, in particular light, means that not all plants are able to survive to physiological maturity. Therefore, the actual plant population may decrease substantially below the initial planting density. This phenomenon is known as 'self-thinning'. The inter- and intra-plant factors that cause one plant to survive at the expense of its neighbour are complex and in many cases poorly understood. However, it is rare for self-thinning to occur at any of the typical planting densities used in tropical agriculture.

Yield

For many crops, the shape of the relation between reproductive yield and planting density may be similar in many respects to that for total dry matter. This is particularly the case in determinate crops such as cereals. However, there are certain factors that distinguish the yield/planting density relation, particularly at high planting densities.

The shape of this relation, again when water supply to roots is not limiting, is also shown in Fig. 6.1. Here, the yield plateau occurs at N3 a population that is similar to that for maximum total dry weight. This is because the ratio of reproductive dry weight, G, to total dry weight, W, i.e. the harvest index, H, often remains conservative across a wide range of populations and growing conditions. In these circumstances, $G = HW$, and the expressions in Equation 6.1 can be used to describe curves for both total dry weight and reproductive yield against population and the shape of these curves will be similar. However, at high planting densities, the conservative nature of H breaks down because, as plant population increases, each additional plant is able to allocate proportionately less and less to reproductive yield. Eventually, the population of plants is so great that each plant simply has enough resources to produce a minimum vegetative weight, w_0, with nothing available to allocate to grain. The planting densities at which H begins to decline and eventually reaches zero are indicated in Fig. 6.1 by N4 and N5 respectively.

Rectangularity

Plant population responses can be considered both in terms of the number of plants per unit area of ground (i.e. planting density) and the spatial arrangement of plants in relation to each other (i.e. planting rectangularity). Because few studies differentiate between density and rectangularity effects on crop productivity, it is often difficult to identify the exact influence of each on the relationship shown in Equation 6.1. Although there is evidence that total productivity is relatively unaffected by the rectangularity of planting (Bell *et al.*, 1987; Gardner and Auma, 1989) it can influence the reproductive yields of some crops (Bell *et al.*, 1987). This is probably because competitive effects become acute later in the season during the period when crops are allocating resources to grain yield.

6.1.2 Water moderately limiting

So far, this discussion has concentrated on ideal circumstances where crops are growing in conditions of plentiful water throughout the season. However, in most cases crops experience soil water deficits at least at some stage in their lives. Clearly, there are many degrees of water deficit, both in terms of the timing and the extent of the stress that a crop experiences. However, to illustrate the general principle of how water shortage affects the relation between crop productivity and planting density, we can present two examples in which water is: (i) moderately limiting and (ii) extremely limiting for crop productivity.

Total dry weight

In Fig. 6.1, the plateau of maximum crop productivity was determined by available radiation and this value could be achieved as long as transpiration could match the reference evaporation (E_{to}) set largely by the atmosphere. However, in dry, tropical regions, the rate of water uptake by crops is often substantially less than that required to meet E_{to} and drought may restrict crop growth for much, if not all, of the season. As a result, the initially linear slope of the relation between dry matter and population is curtailed before N0 (Fig. 6.2) because transpiration is now determined by the supply of water to roots and not by the potential demand of the atmosphere. In practice, maximum productivity (W_m) depends on *actual evaporation* (E_M) during the season which is set by the ability of plant roots to access the store of available water in the soil. Because the supply of water is less than that required by each plant to achieve its genetic potential, the optimum population set by W_m (i.e. N6) is less than the comparable population (N1) where water supply can meet atmospheric demand. Similarly, competition for water means that there may even be a decline in total productivity at higher planting densities. This is because, whilst plants at an intermediate population were all able to achieve a finite weight set by E_M, at the higher density some plants may have used

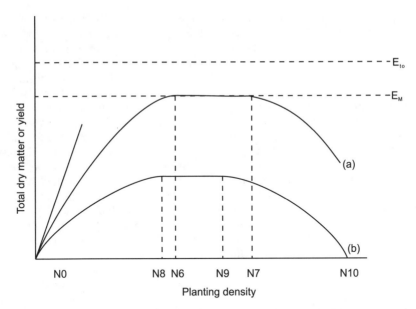

Fig. 6.2. The relation between total dry weight (a) and yield (b) of a crop with increasing planting density when water is moderately limiting. N6, planting density at which total dry weight reaches a plateau; N7, planting density at which crop dry weight begins to decline; N8, planting density at which yield reaches a plateau; N9, planting density at which yield begins to decline, N10, planting density at which yield declines to zero; E_{to}, Reference evaporation; E_M, maximum available soil moisture.

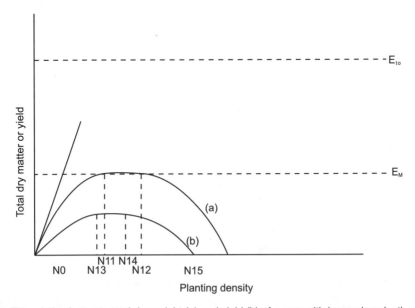

Fig. 6.3. The relation between total dry weight (a) and yield (b) of a crop with increasing planting density when water is severely limiting. N11, planting density at which total dry weight reaches a plateau; N12, planting density at which crop dry weight begins to decline; N13, planting density at which yield reaches a plateau; N14, planting density at which yield begins to decline; N15, planting density at which yield declines to zero; E_{to}, reference evaporation; E_M, maximum available soil moisture.

water early in the season but failed to produce any biomass at final harvest because of severe water stress later in the season (Fig. 6.2, N7).

Yield

When water is limiting, the consequences for grain yield are more evident than in the case of total biomass. Figure 6.2 also shows the theoretical shape of the yield/population relation for moderately limiting soil moisture. Here, the population at which the curvilinear phase ends (N8) is again lower than the comparable values for total biomass. Similarly, the populations at which yield begins to decrease (N9) and eventually reaches zero (N10) are lower than those for total biomass.

6.1.3 Water extremely limiting

In both the above cases, the supply of water was sufficient for crops to achieve their maximum productivity and yield across a fairly wide range of planting densities. However, in many climates, the supply of water to plant roots is so limited that a plateau for biomass and yield is rarely, if ever, achieved. In these circumstances, a knowledge of the optimum planting density is critical. The case for extremely limiting soil moisture is shown in Fig. 6.3. Here, there is little evidence of an asymptote even for total biomass and, in practice, there is a parabolic relation between both total dry weight and reproductive yield against planting density. The appropriate populations for optimum biomass and yield are indicated by N11 and N13, respectively, and the comparable populations where biomass and yield decline are similarly indicated by N12 and N14. The population where yield declines to zero is N15.

Figure 6.4 combines the population responses for total dry matter and yield described in Figs 6.1–6.3. The solid lines indicate the optimum population ranges for total dry matter under the unstressed (N1–N2), moderately unstressed (N6–N7) and severely stressed (N11–N12) conditions described earlier. The corresponding dashed lines indicate the corresponding population ranges for yield under

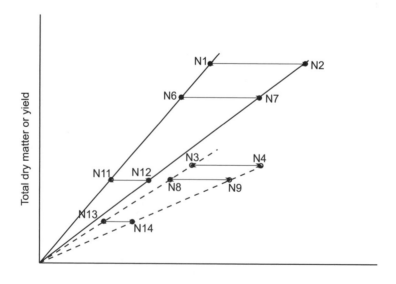

Fig. 6.4. The agronomic ranges of maximum total dry matter (N1–N2, N6–N7, N11–N12) and yield (N3–N4, N8–N9, N13–N14) in response to the unstressed, moderately stressed and severely stressed examples presented in Figs 6.1–6.3.

unstressed (N3–N4), moderately stressed (N8–N9) and severely stressed (N13–N14) conditions. The diagram shows that not only is there a reduction in optimum population with water stress but also that the population range for maximum yield is narrower than that for total dry matter. In particular, the margin for error in selecting the best population declines dramatically with increasing stress.

6.1.4 Predicting the optimum planting density

The above relations show the importance of selecting the most appropriate planting density for any particular crop, which matches the likely supply of resources available during the season. Clearly, where rainfall is plentiful and/or irrigation is available, selecting a precise planting density is not essential and the optimum total dry weight and yield can be achieved across a wide range of plant populations. However, where water is a limiting factor the choice of an appropriate planting density is important and in extreme cases may be critical. The question remains: 'how can we use the theoretical and idealized relations shown in Figs 6.1–6.4 to calculate the most appropriate planting density for a particular crop at any specified location?'

One approach is to make use of the fundamental link that exists between the amount of dry matter accumulated by a crop and its transpiration over the same period. In Chapter 3, the concept of a crop-specific 'transpiration equivalent of dry matter' (Ω_w; g kPa kg^{-1}) was introduced. A knowledge of Ω_w for any particular crop means that, at least in principle, an estimate of seasonal transpiration can provide a means of predicting total crop biomass. In other words, the axes labelled 'total dry matter' in Figs 6.1–6.4 can be replaced by ones labelled 'transpiration'. So, if the amount of water lost directly from the soil surface, E_s, can be measured or estimated, a knowledge of the seasonal evaporation, E_t, at any site can be used to predict the total dry matter, W', of a crop from

$$W' = \frac{\Omega_w E_p}{D} \quad (6.2)$$

where D is the mean seasonal saturation deficit and E_p is total transpiration.

In practice, it is difficult to measure E_p directly or to estimate E_s and hence deduce E_p from $(E_t - E_s)$. However, as a first approximation, when evaporation is limited by the supply of water and E_s is small, E_p can be assumed to be equal to a storage term, S, which depends on the rainfall, R, that infiltrates the soil during the season, plus any applied irrigation, I, and any water already available in the soil profile at sowing, that is

$$W'_s = \frac{\Omega_w S}{D} \quad (6.3)$$

Equation 6.3 assumes that any rainfall plus irrigation in excess of the water-holding capacity of the soil is lost as runoff and/or drainage and that all the remaining water is used in transpiration. For crops where S can meet the seasonal demand of the atmosphere for water, and if E_s remains small, E_p is close to E_o, that is

$$W'E_o = W_m \frac{\Omega_w E_o}{D} \quad (6.4)$$

Thus, reproductive yield can be calculated from either,

$$G'_s = H \frac{\Omega_w S}{D} \quad (6.5)$$

when water is limiting, or from

$$G'E_o = H \frac{\Omega_w E_o}{D} \quad (6.6)$$

when water is non-limiting.

The above approach has rarely been tested in practice. However, Azam-Ali *et al.* (1993) demonstrated the validity of the method for predicting the total productivity and optimum plant populations for groundnut crops growing at different sites in India. At the simplest level, the method can be used to provide a rough-and-ready calculation of the likely productivity of a particular crop for a given site and season. Based on this calculation, a range of planting densities can be carefully selected to span the cardinal populations identified in whichever of Figs 6.1–6.3 most closely resembles the growing conditions expected for any particular site or season.

This last point has important practical implications and illustrates how a mechanistic approach can aid rather than displace conventional agronomic experiments. At field stations throughout the world, plant population experiments are notoriously repetitive and time consuming and impose demands on space and resources that could be better allocated to other investigations. By using the approach described above, the optimum population for a crop at any particular site can be rapidly determined by identifying the range over which experimental effort should be concentrated on a minimal number of representative planting densities. This 'combined' approach is particularly time- and cost-effective when a new crop is being introduced to a particular site or region or when a previously limiting resource, such as water, can be ameliorated through the provision of irrigation with consequences for other agronomic practices.

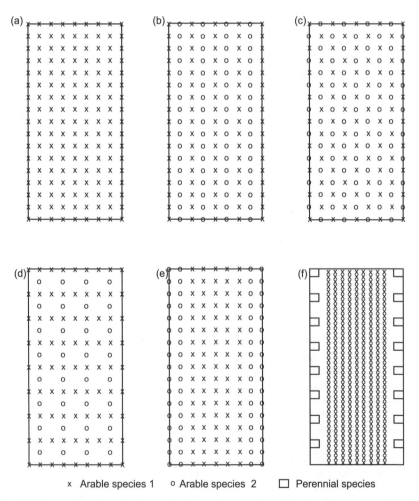

x Arable species 1   o Arable species 2   ☐ Perennial species

Fig. 6.5. Spatial arrangements of various mixed cropping systems. (a) sole crop composed of a single species; (b) row intercrop composed of two species; (c) mixed intercrop composed of two species; (d) additive intercrop; (e) strip intercrop; (f) agroforestry.

6.2 Mixed cropping

In many parts of the tropics, it is more usual for farmers to grow mixtures of different species together rather than single species alone. These mixtures may range from simple combinations of two or more annual crops through to complex arrangements where annual crops are associated with perennial crop and/or tree species.

Before considering why farmers choose to grow mixtures of species and how these might be analysed in relation to their capture and use of resources, it is useful to make some definitions of various types of vegetation in agricultural systems. Table 6.1 summarizes some of the different forms of vegetation that exist in tropical agriculture and Fig. 6.5 graphically illustrates some of these mixed systems in terms of the spatial arrangement of individual plants.

The terms and definitions presented in Table 6.1 are not exhaustive but serve to illustrate a number of important features. First, there is a general increase in biologi-cal complexity from monocrops, where very similar plants are grown alongside each other simultaneously and sequentially from one season to the next, through to the multiplicity of agroforestry systems where diverse types of plants exist together within and between seasons. Second, in temperate, intensive systems of agriculture, most crops are grown as sole crops where management decisions are targeted towards the average needs of a particular species or even variety. As we move into lower input systems, particularly in the tropics, there is an increasing likelihood of biologically complex cropping patterns in relation to the combinations of species grown. This tendency may become more evident as we move from relatively benign to more hostile environments in terms of the availability of natural resources, especially water. This means that, in many dry tropical regions, mixed cropping is the rule rather than the exception.

Despite their popularity with growers, until recently there has been a remarkable lack of enthusiasm shown in advocating and

Table 6.1. Definitions of different forms of agricultural vegetation.

Sole crop	One crop variety grown alone in a pure stand in a single season
Monocrop	The repetitive growing of the same sole crop on the same piece of land
Intercrop	Growing two or more species simultaneously in the same field for at least part of the life of each species
Blend	Growing a mixture of varieties of the same species together on the same piece of land
Mixed intercrop	Growing two or more species simultaneously on the same piece of land with no distinct row arrangement
Row intercrop	Growing two or more species simultaneously where one or more species is planted in rows
Additive intercrop	An intercrop where plants of one or more species are sown in addition to the normal population of a sole crop
Replacement intercrop	An intercrop where the population of plants is identical to that of the component sole crops
Relay cropping	Growing two or more species simultaneously during part of the life cycle of each species
Strip intercrop	Growing two or more species simultaneously in different strips wide enough to permit independent cultivation but narrow enough for the species to interact
Agroforestry	Land-use systems in which trees or shrubs are grown in association with crops or pastures

improving such systems by researchers and extension agencies. The reasons for this ambivalence are diverse but, no doubt, the complexity and range of permutations that exist in various types of crop mixtures has daunted all but the most dedicated researchers. The purpose of this section is to assess the basic dynamics of intercrops in terms of their capture and use of resources and to evaluate methods by which their performance can be compared with that of their constituent species grown as sole crops.

6.2.1 Intercropping

The most common type of mixed crops are intercrops and, not surprisingly, the bulk of research studies on mixed cropping have been completed on various forms of intercrops. Therefore, the discussion that follows will predominantly refer to examples from intercropping. However, it is important to be aware that the principles that govern the behaviour of intercrops can usefully be applied to other forms of mixed cropping and there are a number of recent studies that have extended general principles to interactions between trees and arable species (see, for example, Ong and Huxley, 1996).

6.2.2 Why intercrop?

The reasons why farmers in the tropics choose to grow intercrops are complex and diverse. Often, scientists argue that by growing two or more species together as intercrops there is an improvement in the overall yield of the system than when the same species are grown alone. However, the reasons why farmers decide to grow intercrops often have little or nothing to do with the *possibility* that intercrops provide greater combined yields *per se* than their constituent sole crops. Often, intercropping is chosen as an insurance against the risk of complete crop failure or at least to ensure a greater stability of crop yields in variable environments. Usually, the more variable the environment the greater is the annual variability in crop yields. Often, this variability is reduced when crops are grown together because different species are not equally affected by environmental, pest or disease constraints. The reduced impact of pests and diseases may occur because specialized pathogens of one species may settle on non-host components of the intercrop. This so-called 'fly-paper effect' may be enhanced through the additional compensatory growth of the unattacked species.

Intercrops composed of species with different times to full canopy cover may provide greater competition to the establishment of weeds. Alternatively, we might consider each additional species in an intercrop as a 'desirable weed' which makes a positive contribution to final yield which is greater than that lost by other species through competition for resources. Intercrops may also help to provide increased physical protection of the soil against erosion and wind damage. In these cases, the more continuous ground cover for a greater proportion of the season may stabilize the soil microclimate either by reducing windspeeds at or near the soil surface and/or by binding soil aggregates.

As well as these and many other tangible management benefits, there is also the possibility that certain combinations of intercrops may provide a greater total productivity or seed yield than their constituent sole crops. There is evidence that the combined yield of species grown as intercrops may exceed that of their component species grown as sole crops under similar conditions (see, for example, Willey, 1979; Willey and Rao, 1981; Ahmed and Rao, 1982). Reported yield advantages and the methods for calculating them vary but, for example, intercrops of sorghum and groundnut have typically shown yield advantages of between 25 and 40% (Willey and Osiru, 1972; Wahau and Miller, 1978). Davis and Garcia (1983) reported average yield advantages of between 30 and 72% when a maize cultivar was intercropped with two contrasting groups of bean cultivars. It is important that we distinguish between any effect that intercropping may have on the overall production of biomass by an intercrop with that of any changes in reproductive yields.

Often, it is not biomass *per se* that is affected by intercropping but the combined reproductive yields of each component species within an intercrop. This yield improvement is most evident in the case of cereal/legume intercrops where an improvement in the harvest index of the cereal component may more than compensate for the slight yield loss incurred by the legume fraction (see, for example, Natarajan and Willey, 1980a, b; Trenbath, 1986; Azam-Ali *et al.*, 1990; Ong, 1991; Azam-Ali, 1995). However, it is not always possible to contrast the relative changes in total biomass with those of reproductive yields, since many studies do not present data for both fractions.

6.2.3 What to intercrop?

The choice of species that are selected for intercropping varies with geographical location and local preference. However, the actual combinations of particular crops are determined primarily by the length of the growing season, the relative morphology of each species and the physiological adaptation of crops to particular environments. In areas of low rainfall, early maturing and drought-tolerant species such as millet and sorghum usually predominate. In areas of higher rainfall, various combinations of cereals and legumes with different maturities are used. Combinations of rice and other cereals or legumes may be found in higher rainfall areas with a single extended rainy season such as South-east Asia (Ruthenburg, 1980). Ofori and Stern (1987) summarize the typical distributions of various intercrop systems in contrasting climates of the tropics.

For subsistence farmers and their families, obtaining the correct dietary balance of protein and energy from their own crops is important. Therefore, there may be a nutritional aspect to the choice of crops that are grown by the farmer and this often means that cereal and legume species are chosen for cultivation. Where intercropping is the dominant cropping system, this nutritional consideration is expressed in the choice of crop combinations for intercropping.

Despite their potential benefits, the biological and physical complexity of intercrops has deterred scientists from analysing their productivity in terms of their capture and use of resources. Also, our understanding of the overall influence of each environmental factor across different intercrop combinations is often based on field experiments in which the capture or use of only a single resource, e.g. light, is measured. These studies often provide contrasting and often contradictory evidence as to the source or magnitude of any intercrop advantage. Much of the confusion occurs because: (i) there are many different criteria for measuring the performance of intercrops, and (ii) the task of separating the capture and use of individual resources in cropping systems which, by definition, are intimately related is experimentally and mathematically difficult.

6.2.4 Competition and intercrops

Even when a crop is composed of plants of the same species, the processes that determine how individual plants compete for resources are complex. In all but the simplest systems, plants rarely grow without competition from their neighbours for more than a short time after emergence. This is because farmers usually adjust planting density to maximize the productivity of a *community* of plants per unit area of ground rather than that per *individual* in the population (see Section 6.1).

For crops composed of a single species, there are three major considerations when analysing the factors that control their demand for resources. First, plants grow in two media, soil and air, in which different competitive forces exist between the ability of roots to scavenge for nutrients and water and leaves to intercept light and absorb CO_2. Second, plants compete in two dimensions, space and time, and in terms of seasonal growth the ability of individual plants to occupy a given volume of space has implications for their use of resources at different developmental stages in the life of the crop. Third, crops are primarily grown for a finite fraction of their total productivity, usually the reproductive component. The relationship between reproductive yield and total dry matter is not always simple and in many (often indeterminate) species may vary with environment and crop duration.

Despite the complexity that exists even in sole crops, experimenters and, more recently crop modellers, have made much progress in understanding and predicting the behaviour of single crops in terms of their competition for the capture and use of available resources. However, until recently, similar principles have rarely been extended to intercrops which are often grown in low-input systems of cultivation in variable environments and with differing durations for each component species. A feature of intercrops is that for much of their lives plants might compete for resources with neighbours of both the same *and* at least one different species. This competition occurs in both space and time and has different characteristics above and below ground. So, it is appropriate to consider first the competitive factors that distinguish the behaviour of intercrops in relation to their above and below ground environments.

In intercrops, *intra* (i.e. within) species competition usually precedes *inter* (i.e. between) species with regard to competition for light. The early life of a plant within an intercrop is therefore similar to that of a comparable sole crop. However, where there are marked differences in the final height and growth rates of adjacent plants, for example a low-lying legume growing in a row adjacent to a taller cereal, competition between plants may occur earlier than in the comparable sole crop because of the faster and/or taller growing habit of the cereal. This effect can be influenced and exaggerated by planting geometry, row direction, the aspect and slope of the field and latitude. In many circumstances, for the shorter species in an intercrop, interspecific competition for light is a fact of life earlier than comparable inter-row competition in the sole crop. Of course, for the taller species in an intercrop the converse is the case, since adjacent rows in the sole crop will begin to compete earlier in the season than in the intercrop

because each row is composed of plants of similar size and vigour. In a typical cereal/legume intercrop, individual legume plants would therefore be more 'crowded' earlier in the season than comparable plants in the sole crop and individual cereal plants in the intercrop would be less crowded than their counterparts in the comparable sole crop.

In addition to factors such as variations in air temperature, humidity, windspeed and CO_2, the competition that exists within intercrops can be ascribed to the relative capture and use of light, water and nutrients by each component species of an intercrop and by the whole intercrop compared with it sole counterparts.

6.2.5 Light and intercrops

Capture systems

Differences in the competitive abilities of each species within an intercrop have implications for the performance of the whole system. However, understanding the behaviour of each component is limited by our ability to quantify precisely its capture and use of resources. In principle, the capture of light by intercrops can be measured in the same way as that in sole crops, for example, by the use of tube solarimeters (Szeicz, 1974). By placing arrays of solarimeters at ground level across a representative sample of an intercrop, the average transmission of radiation through the whole canopy can be measured. In this way, average values of fractional interception, f, can be compared between the intercrop and each of its sole components. However, this method tells us nothing about the *relative* capture of radiation by each component within the intercrop. Partitioning the amount of light received by each species in an intercrop is experimentally difficult and is prone to large sampling errors. The path and extinction of solar radiation within a stand depends on the location in the horizontal plane where light enters the canopy which, in turn, depends on the position of each species within the canopy, the solar angle and row orientation. In particular, spatial variations in leaf orientation have major implications for radiation distribution within an intercrop. In sole crops, the idealized pattern of light distribution is described by vertical, or erectophile, leaves in the uppermost layers and planophile, or horizontal, leaves at the bottom of the canopy (Duncan, 1971; Bonhomme, 1993). Similarly, the shading effect of a vertically dominant species in an intercrop can be partly reduced if the taller species has erectophile leaves (Trenbath, 1986; Sinoquet and Caldwell, 1995) above an understorey species with predominantly planophile leaves. This arrangement is best exemplified by intercrops composed of a tall 'C_4' species such as millet, sorghum and maize, overlying a 'C_3' legume such as cowpea, common bean or groundnut. In such cases, higher intensities of radiation can be usefully intercepted by the cereal component, allowing the penetration of lower intensities that are below the saturating irradiance for C_3 species. However, this is the idealized case and actual leaf orientation is not simply a function of the species in question. Any factors that influence the supply and demand for resources, such as competition for light or water, may influence the way that individual plants distribute their foliage.

Attempts at estimating radiation intercepted by each species in an intercrop are rare. Marshall and Willey (1983) divided the radiation within a millet/groundnut intercrop into that captured by each species using quantum sensors above and below the sole crop and intercrop canopies. Measurements made at intervals below the canopies were used to calculate variations in the transmission of *photosynthetically active radiation* (PAR). Expressed per row, the millet intercepted 112% more PAR in the intercrop than in its equivalent sole crop. In contrast, groundnut intercepted 27% less PAR than in its comparable sole crop. Azam-Ali *et al.* (1990) found a similar pattern of increased capture of radiation by the cereal component and decreased capture by the legume component of a sorghum/groundnut intercrop.

Partitioning of light from radiation

measurements is only feasible when component species are segregated into distinct volumes, such as where there is a distinct vertical stratification, e.g. shrub/grass mixtures (Tournebize and Sinoquet, 1995) or horizontal stratification as in the row intercropping examples mentioned earlier for millet/groundnut (Marshall and Willey, 1983) or sorghum/groundnut (Azam-Ali *et al.*, 1990). As with sole crops, most analyses of intercrops apply various forms of Beer's law (see Chapter 2) to the penetration or interception of radiation within replacement intercrops (i.e. where there is no increase in the overall population of plants). In these cases, extinction coefficients (k values; see Chapter 2) are usually derived empirically to provide a radiative balance for each species in a particular intercrop combination. However, it is unclear whether the k value for a species growing within an intercrop remains similar to that for the same species grown as a sole crop. In fact, Sivakumar and Virmani (1984) found little difference in k between sole maize (0.64), sole pigeon pea (0.69) and a maize/pigeon pea intercrop (0.66) in which two-thirds of the leaf area was produced by maize. Unfortunately, there is no widely accepted method of calculating the k value of any particular species within different forms of intercrop, nor of comparing this with its typical value within a sole stand. Given that we have not as yet overcome the difficulties of describing light attenuation within heterogeneous populations of sole crops, the task of providing general principles to describe the behaviour of radiation in intercrop canopies is still at a rudimentary stage.

A modified form of Beer's Law can perhaps be most usefully applied to *additive* intercrops, i.e. those which result in an increase in total plant population through the addition of one or more species to an existing population of a first species sown at the same density as its sole counterpart. Here, the canopy structure can be considered as horizontally homogeneous but vertically heterogeneous (Ong *et al.*, 1996). The intercrop canopy can be stratified into several horizontal layers such that

$$S_1 = S(1 - e(k_{1A}L_{1A} + k_{1B}L_{1B}))\ (6.7)$$

where S is the radiation reaching canopy layer, l, S_1 is the quantity of radiation intercepted by layer l and k_{1A}, k_{1B} and L_{1A} and L_{1B} are the extinction coefficients and leaf area indices for layer l in species A and B respectively. Now, from Keating and Carberry (1993), interception by species A in layer l (S_{1A}) can be calculated as

$$S_{1A} = S_1 \frac{k_{1A}L_{1A}}{k_{1A}L_{1A} + k_{1B}L_{1B}} \qquad (6.8)$$

After calculating similar estimates for species B, total interception for the intercrop and each species can be obtained by summing the interception values for all canopy layers. One of the assumptions underlying this modification is that the increase in planting density means that the intercrop stand is a closer approximation to the randomly distributed foliage required for Beer's Law than its sole counterparts. However, the above example depends largely on the assumption without the benefit of accurate measurement. Again, we do not know the exact effect of competition from one or more additional species and the increase in overall planting density on the actual orientation and arrangement of leaves within various forms of additive intercrop.

Nevertheless, if species are mixed within the same space, as in the additive example mentioned above, measurements of radiation are unable to provide estimates of the light intercepted by each component species unless assumptions are made. For example, Ryel et al. (1990) inserted a photodiode line within an intercrop canopy and estimated proportional light attenuation by each component after assuming a random leaf dispersion, irrespective of species. Sinoquet and Caldwell (1995) have summarized recent developments in techniques to measure light partitioning in intercrops. These include photography linked with digital image analysis to recreate a computerized view of the intercrop canopy. However, until such techniques gain wide acceptance, our understanding of light distribution in mixed vegetation canopies depends largely on various forms of radiation models (see, for example, Lemeur and Blad, 1974; Goel, 1988; Sinoquet and Caldwell, 1995).

A further complication when analysing the seasonal interception of radiation by intercrops is that most are composed of species with differing durations. In such cases, we need to account for a temporal element of reduced competition. In the case of cereal/legume intercrops, the cereal component may be harvested well before the legume. This means that the legume rows in the intercrop will continue to grow during much of their reproductive phase without competition from the taller cereal. However, this does not mean that the light interception for the legume component remaining in the 'intercrop' will be similar to that in a comparable sole crop because there are now a number of missing rows which do not compete for solar radiation. Here again, methods for estimating radiation interception by sole crops made up of non-random or clumped foliage must be used. However, the nature of radiation interception by a residual legume species in an intercrop may not even be the same as a sole crop grown in the same discontinuous arrangement. This is because the previous history of competition from a taller and more aggressive neighbouring species in the intercrop has implications for the shape and extent of foliage in the legume on removal of the cereal (Azam-Ali, 1995).

Conversion systems

Not surprisingly, calculating the conversion efficiency of solar radiation for each component of an intercrop is far more complicated than that for sole crops. Where the canopy volume is horizontally stratified, as in the case of the millet/groundnut intercrop mentioned earlier (Marshall and Willey, 1983), it is possible to use measurements of fractional interception and growth to calculate the value of ε_s for each species in the intercrop compared with its sole counterpart. In their example, Marshall and Willey explained the overall advantage of the intercrop through a 46% increase in the conversion efficiency of the groundnut component combined with a 10% increase in the capture efficiency of the millet component with virtually no change in its conversion efficiency. However, such examples are rare and, in most cases, the mean conversion efficiency of the whole intercrop canopy is calculated simply by dividing the total dry matter of all the components by the total amount of solar radiation captured by the complete system. Seasonal mean values for intercrops are presented by Reddy and Willey (1981) for a millet/groundnut intercrop (ε_s = 1.08 g MJ^{-1}) by Harris *et al.* (1987) for a sorghum/groundnut intercrop (for an irrigated treatment ε_s = 1.02 g MJ^{-1} and for a rainfed treatment ε_s = 0.94 g MJ^{-1}).

As a basis for comparison within and between systems, gross values for the whole intercrop should be treated with caution. First, the values tell us nothing about changes in conversion efficiency, either during the period when the component species are competing for resources or for the period after the harvest of part of the intercrop when companion species are still growing. Second, there are virtually no reports of the contribution that the weight of roots of each species make to the overall conversion efficiency of the system. Again, discrepancies in the calculated values of ε_s for different intercrops are likely to result from this omission, especially where the relative contribution of roots is different between component species. Third, there appear to be few published values of ε_s for intercrops where the energy equivalent of the reproductive fraction has been used to adjust total biomass. This has important implications: (i) where component species differ in the lipid fraction, e.g. cereals and oilseeds and (ii) where the practice of intercropping changes the ratio of reproductive (or economic) yield to total dry weight, i.e. the harvest index, *H*, of one or more components.

Therefore, to calculate the true seasonal conversion efficiency of a crop system we must divide the total dry weight of grain, leaves, stems, roots, etc. at final harvest by the total amount of radiation intercepted during the season. Because not all plant organs have the same energy content per unit of dry weight, we need to account for the energy value (MJ g^{-1}) of the lipid, protein and carbohydrate fractions as discussed in Chapters 2 and 3. In this way, the total biological output of a sole crop or intercrop system can be calculated in terms of megajoules, i.e. the same unit used to calculate the seasonal interception of radiation. We are not aware of any study that has successfully partitioned both the radiation captured by each component of an intercrop and its energy-adjusted total dry weight with comparable measurements of each constituent species grown as a sole crop.

6.2.6 Water and intercrops

Capture systems

Allocating the amount of water captured by the root systems of each component in an intercrop is as problematic as that for light. The degree of interaction between the roots of adjacent plants will depend on their morphology, lateral extent and depth as well as on the configuration of species within an intercrop. An added complication is that the root systems of many species are extremely dynamic and their dimensions can be modified by the relative distribution of resources, such as water and nutrients, within the soil profile. An important additional factor in many tropical soils is that the extent and distribution of roots is constrained by the structure and depth of the soil profile. Although similar constraints may apply to the root systems of sole crops, the *relative* speed and ability of roots of different species to extend into soils of variable structure and depth has implications for the amount of water that each species can extract from a given soil reservoir. Of course, roots do not grow in isolation from the rest of the plant and differences in water uptake between each component of an intercrop may in turn alter the canopy geometry and assimilation rate of the above ground elements.

In theory, intercropping facilitates a more complementary use of water than in sole crops where each plant is competing for water at the same time and in the same space as its neighbours. The greatest degree of complementarity should, therefore, occur in intercrop combinations where species differ substantially in the shape, extension rate and duration of their root systems. However, although there may be differences in the timing of water use between intercrops and sole crops, experimental evidence suggests that seasonal water use is usually quite similar. Ong *et al.* (1996) summarize data for water use from ten studies where seasonal rainfall ranged from less than 100 mm to over 570 mm. In most cases, water use by the intercrop treatments was within a few per cent of the corresponding sole crops.

There have been few attempts to directly measure the uptake of water by different components within an intercrop. In practice, water uptake by each component species can be calculated either by measuring the transpiration of one component and calculating that from the other species from a knowledge of the soil water balance of the whole system (see Chapter 3), or by measuring the transpiration of each component separately. Various techniques exist to measure transpiration either directly from the foliage of each species, for example via diffusion porometry (see for example, Azam-Ali, 1984; Wallace *et al.*, 1990) or, more recently, using sap flow techniques (see Marshall *et al.*, 1994; Howard *et al.*, 1995). A review of the different methods of estimating or measuring transpiration from each component species is presented by Ong *et al.* (1996).

Conversion systems

The difficulties identified above in partitioning the relative capture of soil moisture by different components of intercrops have consequences for calculating the respective relationships between dry matter production and transpiration for each species. In view of these complexities, the few studies that describe the water use by intercrops and their comparable sole crops generally assume that the dry matter/transpired water ratio (ε_w; g kg^{-1}) for each species remains the same whether it is grown as part of an intercrop or as a sole crop. From this

assumption, total evaporation by an intercrop can be partitioned between its component species from a knowledge of the dry matter produced by each component within the intercrop and the total crop water balance. Where actual measurements of water use have been made, these have usually been confined to comparisons of water uptake between the whole intercrop and that of its constituent sole crops. Again, the assumption has been that the value of ε_w for each species is unaltered by cropping system. To calculate whether the value of ε_w for a species grown in an intercrop is different from that in a sole crop requires measurements of water use and total dry matter production (including roots) by each component. Of course, this has the added complication that intermingling roots of different species must be separated to establish their individual weights.

Despite the complications involved in measuring water capture and conversion and the individual components of biomass, the premise that ε_w remains conservative across various cropping systems is more tenable than the assumption that ε_s is constant. There are several reasons for this. First, we have seen in Chapter 3 that the general relation between dry matter production and transpiration is more robust both during the season and across seasons than that between dry matter production and accumulated light interception, especially under conditions of soil moisture stress. Second, the Marshall and Willey (1983) study shows that the nature of light attenuation in intercrops is likely to influence the value of ε_s in at least one of the component species because the pattern of radiation distribution is radically altered within the canopy volume with consequences for the light response curves of individual leaves. In practice, the most likely influence on dry matter/transpiration relations within intercrops is through the normalizing effect of saturation deficit (D) on ε_w (see Chapter 3). Where intercropping can be shown to influence atmospheric humidity within the canopy volume, these alterations can be used to recalculate the effect of D on seasonal transpiration.

The complexities associated with partitioning each component of the water balance to account for drainage, soil surface evaporation and transpiration from each constituent species has led to contradictory conclusions about possible changes in water use efficiency within intercrops. However, it is likely that any apparent changes in the efficiency of water use by each species can be explained by other consequences of intercropping, such as a reduction in total evaporation from the soil surface. In such cases, water available for transpiration, i.e. the water *capture efficiency* of intercropped species may have been improved rather than the water *conversion efficiency* of each species (Stigter and Baldy, 1995).

6.2.7 Nutrients and intercrops

While the energy flow for solar radiation described in Chapter 2 is a one-way process, nutrients can cycle around an ecosystem within and between seasons. Therefore, the soil type, previous cropping history and the particular combination of species within an intercrop all influence the relative competition between species for soil nutrients. Because nutrients, in particular nitrogen, are dissolved within the soil solution, any changes in the seasonal water balance of an intercrop relative to its sole counterparts will influence the capture of specific nutrients.

A further complication in the nature of nutrient capture in mixed systems is that the rate of uptake in a species varies with plant age, and therefore the period of maximum nutrient demand for one species in an intercrop may not coincide with that of another. The balance between the amount of nutrients captured by each species within mixed crops will also depend on the overall uptake of nutrients by the whole system. Where the rooting depth of each species within an intercrop is similar, total nutrient availability may also be similar. However, this is not always the case. In climates where rainfall is not surplus to soil moisture storage during the season, nutrient supply will precede nutrient demand and timing of demand may

not be critical to the total availability. However, in the humid tropics, where rainfall exceeds seasonal evaporation, products of early mineralization may be washed into deeper soil layers and may, therefore, be leached beyond crop roots. In these circumstances, deep rooted components within mixed systems can act as a 'safety net' by intercepting nutrients, in particular nitrogen, on its way to deeper layers (van Noordwijk *et al.*, 1996). The most extreme example of this deeper mining and uptake of nutrients occurs in agroforestry systems where deep-rooted perennials are able to act as 'nutrient pumps' bringing nutrients up to the soil surface for both the simultaneous and sequential benefit of the whole system.

A subject which has generated much interest is the possible complementary effects of intercropping non-legumes with nitrogen-fixing legumes on the overall nitrogen economy of the whole system. There is evidence that N_2 fixed by the legume can be directly transferred to the associated non-legume via root/nodule exudation (Wacquant *et al.*, 1989; Horst, 1995), and/or via vesicular fungi (Van Kessel *et al.*, 1985) or through more efficient uptake of N released by the legume (Hamel *et al.*, 1991). However, it is generally agreed that these various processes contribute little to the simultaneous nitrogen economy of the non-legume component (Van Kessel and Roskoski, 1988; Reeves, 1992), i.e. there is no advantage to the non-legume in the same season. It is more likely that decomposition of legume residues during the growth cycle provides a residual benefit to any succeeding non-legume crops – an advantage which need not be a consequence of intercropping.

In summary, it seems that a greater uptake of nutrients by intercrops may be a consequence rather than a cause of any increase in their overall biomass (Morris and Garrity, 1993). Any competition that does occur between the different species of an intercrop is more likely to be for mobile nutrients, such as NO_3^-, than for nutrients with low mobility such as K and P. In these latter cases, the comparative root length densities in the intercrop would have to be implausibly greater than those in the sole crops for any measurable changes in relative uptake (Horst, 1995).

Nutrients and light interception

In Chapter 2, it was shown that the relationship between the N content of vegetation and cumulative dry matter is a function of the light microclimate within the stand. Thus, for an isolated plant the decline in N% is usually only due to a decrease in the ratio between metabolic and structural tissues which have different N concentrations. The rate of decline in N% becomes progressively more acute with increasing plant population in sole crops as light is more rapidly attenuated within the canopy. The decline in leaf N% with declining radiation levels within the canopy of an intercrop has also been observed (Soussana and Arregui, 1995). Since light microclimate is highly dependent on planting density, the decline in N% of plant tissue with depth in the canopy is more a function of planting density than cropping system. In the case of intercrops, the 'additive' or 'replacement' nature of the crop combination is probably more important on N distribution than the exact combination of species *per se* (Cruz and Soussana, 1997).

Nutrients and transpiration

In the case of water use, we have seen that any intercrop advantage is probably due to a greater capture of water rather than any changes in transpirational efficiency for each component species. Any increase in the total amount of water captured by the intercrop may be associated with the ability of different species to explore greater depths of soil. However, nutrients, and in particular nitrogen, are mainly localized in the upper layer of the soil and therefore, even where roots are able to penetrate to greater depths and extract more water, this may not necessarily result in a proportional increase in nutrient uptake. Of course, where nutrients, such as nitrogen, are highly mobile, their availability with depth in the soil profile will depend on the balance between annual rainfall and soil moisture storage and

this may influence the relative balance between transpiration and nutrient uptake.

6.3 Calculating the performance of intercrops

The obvious basis for comparing the response of different sole crops to a particular agronomic practice is to measure their respective yields with and without the intervention. Similarly, we can attempt to explain the causes of any differences in total biomass and yield between sole crops through measurements or estimates of their capture and conversion of resources such as solar radiation and water. For intercrops, however, the situation is far more complex. In this section, we consider the issues involved in assessing the performance and resource use of intercrops and how these might be analysed relative to the performance of the constituent species grown as sole crops.

To provide a rational basis for quantifying the capture and use of resources by intercrops and their component sole crops, we need to use a common method to evaluate the performance of each component of an intercrop and its constituent sole crops. In contrast to the dearth of experimental information on resource capture and use by intercrops, the scientific literature is replete with a plethora of numerical indices of intercrop performance. These are usually based on measurements of crop biomass or yield and rarely provide any insights into the causes of any differences that may occur between intercrops and their comparable sole crops. It is not our purpose here to review all the methods that have been advocated for assessing the performance of intercrops. However, it is useful to compare three existing techniques which each illustrate the importance of different aspects in the assessment of intercrop performance. These are the Land Equivalent Ratio (LER), the Area–Time–Equivalency Ratio (ATER) and the Crop Performance Ratio (CPR).

6.3.1 Land Equivalent Ratio (LER)

Willey and Osiru (1972) proposed the concept of the 'Land Equivalent Ratio' (LER) as a simple index defined as 'the relative land area required from sole crops to provide the same yield achieved by an intercrop composed of the same species'. For an intercrop composed of two species, i and j:

$$\text{LER} = (Y_{ij} / Y_{ii}) + (Y_{ji} / Y_{jj}) \quad (6.9)$$

where Y is the yield per unit area, Y_{ii} and Y_{jj} are sole crop yields of the component crops i and j, and Y_{ij} and Y_{ji} are intercrop yields. The partial LER values, L_i and L_j, represent the ratios of the yields of crops i and j when grown as intercrops, relative to their equivalent sole crops. Thus,

$$L_i = (Y_{ij} / Y_{ii}) \quad (6.10)$$

and

$$L_j = (Y_{ji} / Y_{jj}) \quad (6.11)$$

The total LER of an intercrop is simply the sum of its two partial land equivalent ratios (i.e. LER = $L_i + L_j$). When LER = 1, there is no advantage to intercropping and when LER >1 a larger area of land is required to produce the same yield of sole crop of each component than with a comparable intercrop. The LER has been widely used to demonstrate intercropping advantages across a wide range of species combinations and environments (see, for example, Willey, 1985; Ofori and Stern, 1987) and remains the most widely used index for assessing the relative performance of intercrops. However, a major criticism of the LER concept is that it takes no account of the relative durations of each species within the intercrop or sole crop systems. As a result, published estimates of LER often exaggerate intercropping advantages because the land left unused after the harvest of the shorter duration sole crop is not included in the calculations.

6.3.2 Area–Time–Equivalency Ratio (ATER)

To overcome the conceptual weakness (i.e. absence of time) inherent in the LER concept, Hiebsch and McCollum (1987) proposed an 'Area–Time–Equivalency Ratio' (ATER) which defines yield as a function of both land area (a) and time (t). To present ATER mathematically, an expression is needed which includes the duration of land occupied by an intercrop as well as the duration for each component sole crop. This can be done by redefining Y as the quantity q per unit of time, so that $Y = q/at$. An area time equivalency ratio can then be defined as

$$\text{ATER} = \sum_{c=1}^{n} Y^i_c / Y^m_c \quad (6.12)$$

Thus, productivity by species c in the intercrop can be expressed in terms of both area and time (where $Y^i_c = q^i_c/at^i$, where t^i is the intercrop duration). Similarly, the productivity of species c as a sole crop can also be expressed as a function of area and time ($Y^m_c = q^m_c/at^{mi}$, where t^{mc} is the sole crop duration). Equation 6.12 can then be expanded and rearranged to give:

$$\text{ATER} = \sum_{c=1}^{n} (q^i_c / at^i)/(q^m_c / at^m_c) \quad (6.13)$$

and

$$\text{ATER} = \sum_{c=1}^{n} \frac{(at^m_c)(q^i_c / q^m_c)}{at^i} \quad (6.14)$$

For each crop c, $(at^m_c)(q^i_c/q^m_c)$ is the area × time, i.e. hectare days (ha day) required by sole crops to produce the quantity of c achieved in the intercrop. Where this quantity is calculated for all species in the intercrop, ATER is the ratio of area × time required in sole cropping to area × time used by the intercrop to produce the same quantities of all the component crops, that is

$$\text{ATER} = \sum_{c=1}^{n} at^m_c / at^i = at^m / at^i \quad (6.15)$$

ATER can now be compared in a computational form that is analogous to LER, that is

ATER $= [(y^i_1/y^m_1)(t^m_1) + (y^i_2/y^m_2)(t^m_2) +$
$...(y^i_n/y^m_n)(t^m_n)]/t^i$

$$(6.16)$$

When time is included in the calculation, Hiebsch and McCollum (1987) observed that 'large land-use advantages ascribed to growing food crops in mixtures disappeared'. They concluded that most intercrops use land and area with the same efficiency as pure stands of the same species.

6.3.3 Crop Performance Ratio (CPR)

Irrespective of whether intercropping produces an advantage or not, in their conventional forms neither LER nor ATER identify the physiological or physical processes responsible for any differences that may occur between intercrops and sole crops. Furthermore, neither index presents the absolute or relative biological efficiencies of the systems in terms of the amount of biomass or yield fixed relative to the energy captured during the season. This omission may be acceptable in sole crop systems where there are unlikely to be major changes in the energy values of plant products across seasons or treatments. However, a knowledge of energy equivalents is important in intercrop systems which are often composed of plant products with substantially different composition. By referring all calculations to the energy equivalent of biomass (MJ) we can use a 'common currency' with which to compare biological outputs of different species composed of organs, each with different energy values.

To examine the processes responsible for differences between the performance of intercrops and their constituent sole crops, we therefore require a method of analysis which:

- calculates the total productivity of the intercrop (adjusted for energy) per unit area of land relative to that which would have been produced from the same land area by sole crops sown in the same proportions as the intercrop;

- accounts for the effect of differences in crop duration for each species on the performance of the intercrop and sole crops;
- allows the capture or conversion of resources such as solar radiation, water or nutrients by intercrops and sole crops to be compared on the same basis as that for dry matter or yield.

To compare the biological output of an intercrop with that of its component sole crops, we can use the concept of a 'Crop Performance Ratio' (CPR). The use of CPR to assess the performance of sorghum/groundnut intercrops was demonstrated by Harris et al. (1987) and Azam-Ali et al. (1990).

For an intercrop composed of two species (a and b), the Crop Performance Ratio CPR_a for species a is given by

$$CPR_a = \frac{Q_{ia}}{P_{ia}Q_{sa}} \qquad (6.17)$$

where, Q_{ia} and Q_{sa} are its productivity per unit area (MJ m^{-2}) in the intercrop and sole crop, respectively, and P_{ia} is the proportional sown area of species a in the intercrop. A similar calculation can be made for species b so that for both species together:

$$CPR_{ab} = \frac{Q_{ia} + Q_{ib}}{P_{ia}Q_{sa} + P_{ib}Q_{sb}} \qquad (6.18)$$

Because the sole crop values are multiplied by their sown proportions in the intercrop, this provides their 'expected' productivity if unit area of ground had been sown with sole crops in the same proportions as in the intercrop. As with LER a value of CPR >1 indicates an intercrop advantage and a CPR <1 an intercrop disadvantage. The CPR concept can be extended to analyse the capture or use of any resource by an intercrop compared with its constituent species. For example, we can calculate a CPR for intercepted radiation, transpiration, nutrient uptake, etc., in which the expected resource use by an equivalent sole crop is always unity.

For intercrops composed of species with different durations we need to account for the durations, t_a and t_b for species a and b respectively. The duration for each species is assumed to be the same whether it is grown as a sole crop or intercrop and $t_a > t_b$. Then, for the sole crops the rate R (MJ day^{-1}) of productivity is given by

$$R_{sa} = \frac{Q_{sa}}{t_a} \text{ and } R_{sb} = \frac{Q_{sb}}{t_b} \quad (6.19)$$

For the same species in the intercrop, the rates of productivity are given by

$$R_{ia} = \frac{Q_{ia}}{t_a} \text{ and } R_{ib} = \frac{Q_{ib}}{t_a} \quad (6.20)$$

Note that the rate of productivity for species b depends on the duration of species a since land is occupied by the intercrop for this length of time.

Now, the 'time corrected' Crop Performance Ratio (CPRT) for species a is given by

$$CPRT_a = \frac{R_{ia}}{P_{ia}R_{sa}} = CPR_a \quad (6.21)$$

and for species b is

$$CPRT_b = \frac{R_{ib}}{P_{ib}R_{sb}} = CPR_b\left(\frac{t_b}{t_a}\right) \quad (6.22)$$

For the intercrop, CPRT$_{ab}$ is given by

$$CPRT_{ab} = \frac{R_{ia} + R_{ib}}{P_{ia}R_{sa} + P_{ib}R_{sb}} \quad (6.23)$$

$$\Rightarrow \quad CPRT_{ab} = \frac{\left(\dfrac{Q_{ia} + Q_{ib}}{t_a}\right)}{\left(\dfrac{P_{ia} + Q_{sa}}{t_a}\right) + \left(\dfrac{P_{ib} + Q_{sb}}{t_b}\right)}$$

$$(6.24)$$

$$\Rightarrow \quad CPRT_{ab} = \frac{(Q_{ia} + Q_{ib})t_b}{P_{ia}Q_{sa}t_b + P_{ib}Q_{sb}t_a} \quad (6.25)$$

6.3.4 Comparing LER, ATER and CPR

Table 6.2 summarizes the performance of millet/groundnut intercrops and their constituent sole crops grown in the rainy sea-son at the International Crops Research Institute for the Semi-Arid Tropics (ICRISAT), near Hyderabad, India, between 1985 and 1987. The data are taken from Ong *et al.* (1991) and the table was originally presented by Azam-Ali (1995). To calculate the energy equivalents of biomass for total dry matter and yield, the vegetative and reproductive dry weights of millet were multiplied by 17.51 KJ g^{-1} (Passmore and Eastwood, 1986). This calculation assumes that the carbohydrate and protein fractions have the same energy value and that there is no lipid in the seed. For groundnut, where lipids provide a substantial proportion of the seed weight, the reproductive yield was multiplied by 23.41 KJ g^{-1} (McCance and Widdowson, 1978) whilst the vegetative component was multiplied by 17.51 KJ g^{-1}. In addition to total dry matter and final yields, seasonal values of f and ε_s are also presented. Based on these values, sample calculations are presented for CPR, CPRT, ATER and LER. There is insufficient evidence from the data presented in Table 6.2 to make any general conclusions about the validity of the different techniques for presenting the relative intercrop performance. Nevertheless, it is clear that apparent advantages in the intercrop performance expressed in terms of CPR or LER are always reduced in magnitude when adjusted for time by using CPRT or ATER.

6.4 Conclusions

Perhaps the main conclusion that can be drawn from the wealth of empirical and limited amount of quantitative evidence that has been gathered on various forms of intercrop is that they rarely result in greater yields *per se* than their constituent sole crops when factors such as time are taken into account. Expressed on a daily basis, intercropping does not, therefore, appear to be biologically more *efficient* than sole cropping in terms of its capture and conversion of resources. This is not to say that, in practice, intercropping is not worthwhile – the occupancy of a field is usually related to the number of harvests, i.e. crops grown

Table 6.2. Calculation of Crop Performance Ratio (CPR), 'time corrected' Crop Performance Ratio (CPRT), Area–Time–Equivalency Ratio (ATER) and Land Equivalent Ratio (LER) based on values of total dry matter (TDM), reproductive yield, seasonal fractional intercepted radiation (*f*) and conversion coefficient for intercepted radiation (ε_s) for millet/groundnut intercrops grown in central India.

1985	Q_{sb}	Q_{sa}	Q_{ib}	Q_{ia}	$Q_{ia}+Q_{ib}$	CPR	CPRT	ATER	LER
TDM (g m^{-2})	610	280	300	238	538	1.48	1.25	1.19	1.34
TDM (MJ)[a]	10.68	5.45	5.25	4.65	9.90	1.47	1.25	1.20	1.34
Yield (g m^{-2})	123	93	85	82	167	1.66	1.47	1.36	1.57
Yield (MJ)[a]	2.15	2.18	1.49	1.92	3.41	1.57	1.42	1.36	1.57
f (MJ)	584	772			799	1.10	1.01		
ε_s (g MJ^{-1})	1.04	0.36			0.67	1.26	1.04		
ε_s (%)	1.83	0.71			1.24	1.26	1.04		
1986	Q_{sb}	Q_{sa}	Q_{ib}	Q_{ia}	$Q_{ia}+Q_{ib}$	CPR	CPRT	ATER	LER
TDM (g m^{-2})	472	210	227	165	392	1.42	1.23	1.14	1.27
TDM (MJ)[a]	8.26	3.77	3.97	2.92	6.89	1.41	1.22	1.12	1.25
Yield (g m^{-2})	181	16	104	5	109	1.90	1.47	0.73	0.89
Yield (MJ)[a]	3.17	0.37	1.82	0.12	1.94	1.81	1.41	0.73	0.89
f (MJ)	565	473			528	1.06	0.96		
ε_s (g MJ^{-1})	0.84	0.44			0.74	1.37	1.20		
ε_s (%)	1.46	0.80			1.31	1.35	1.19		
1987	Q_{sb}	Q_{sa}	Q_{ib}	Q_{ia}	$Q_{ia}+Q_{ib}$	CPR	CPRT	ATER	LER
TDM (g m^{-2})	647	360	229	263	492	1.14	0.98	0.98	1.08
TDM (MJ)[a]	11.33	7.42	4.01	5.34	9.35	1.11	0.97	0.97	1.07
Yield (g m^{-2})	221	189	87	125	212	1.08	0.96	0.94	1.06
Yield (MJ)[a]	3.87	4.42	1.52	2.93	4.45	1.04	0.94	0.94	1.06
f (MJ)	442	556			614	1.16	1.07		
ε_s (g MJ^{-1})	1.46	0.65			0.80	0.94	0.79		
ε_s (%)	2.56	1.33			1.52	0.93	0.79		

[a] Energy equivalents for TDM, and yield (MJ) were calculated by multiplying the vegetative and reproductive components of sorghum by 17.51 KJ g^{-1} (Passmore and Eastwood, 1986). For groundnut, reproductive yield was multiplied by 23.41 KJ g^{-1} (McCance and Widdowson, 1978) and the vegetative component by 17.51 KJ g^{-1}.

Source: The values for crop productivity for millet in the sole crop (Q_{sb}) and intercrop (Q_{ib}) and groundnut in the sole crop (Q_{sa}) and intercrop (Q_{ia}) and *f* were obtained from data presented by Ong *et al.* (1991) and the original table appears in Azam-Ali (1995).

per year, rather than days in the field and on this basis intercrops consistently outperform sole crops. Similarly, absolute yield is rarely the only criterion of the efficacy of a cropping system, particularly in the seasonally arid tropics. A major consideration is the risks involved in a particular agronomic system. In this regard, intercrops are inherently less risky than sole crops, with checks and balances against catastrophic losses due to predation and disease and the complementary sharing of limited resources.

7

Manipulating Fluxes

Agronomy aims to manage the capture and conversion of environmental resources by communities of arable plants. Fertilizing, weed control and land forming all act to raise or regulate the flow of solar radiation and water through crops. Where a particular resource is available in sufficient quantity and over an appropriate duration to meet the demands of a growing crop, e.g. when seasonal rainfall exceeds potential evaporation, there is little advantage in augmenting the natural supply of this resource. However, the demands of a crop are often greater than the availability of any particular resource and either the supply of each resource must be increased or crop demands must be reduced through agronomic interventions, such as adjusting the planting density or selecting less demanding or better adapted species. As described in Chapter 4, advances in yield have repeatedly been achieved by the judicious combination of plant improvement and agronomy.

Solar radiation is the primary resource for crop growth and the total amount of radiation intercepted during a growing season sets the *potential* maximum growth that can be achieved by any particular crop (see Chapter 2). However, radiation can neither be supplemented in the field nor stored for later use by crops. Therefore, our principle objective here is to consider how the supply of other resources, such as water and nutrients, can be optimized so that the *actual* growth achieved by a crop is as close as possible to its potential maximum. Because

nutrients, either available in the soil or provided as fertilizers, are taken up by plant roots in solution, increasing the availability of water to plant roots has important implications for the supply of nutrients and there is often a close link between water and nutrient uptake by arable crops.

Fertilizing, weed control and land shaping each influence several important processes and by doing so, interact in a complex way. They are nevertheless often treated in experiments as independent factors, and most of the examples given here are unavoidably from studies that relate to only one or other of them. The approach taken, however, is to illustrate common features in the hope that this will encourage future studies to consider flux manipulation as a whole.

7.1 A simple frame of reference

Any management practice can be represented by an input–yield curve, a graph showing how dry matter or yield changes with increasing addition of fertilizer, intensity of weed control or effort spent in soil shaping. These curves have four features (Fig. 7.1).

1. The base value of yield (or biomass) – the performance of the crop in the absence of fertilizer, weed control or soil shaping (A).
2. The 'rate' of increase in yield with incremental additions of material or effort (AB).
3. The plateau at which yield stabilizes at its maximum for the climate (BC).

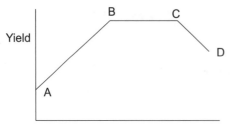

Fig. 7.1. Idealized relation between yield and inputs showing: basal yield, A; the region where yield increases with further inputs, AB; the plateau BC where yield does not respond to inputs; and the decline CD where high inputs reduce yield.

4. In certain instances, a decline in yield at high levels of intervention (CD): this would occur through any toxic effect of high fertilizer dose, through waterlogging or if the presence of an intermediate weed population improved the performance of the crop compared to one on a completely bare field.

This set of straight lines illustrates the biological responses involved. The observed response is generally curved between the lines A, B and C, resulting in a 'diminishing return' as further inputs move the yield ever more gradually towards the plateau.

Fertilizing, weed control and soil shaping are complex subjects requiring their own detailed and specialist treatment. Specific accounts of particular methods are well outside the scope of this book. Instead, we aim to summarize features of the input–response curves, consider their physiological basis and finally touch on broader issues of resource management. The treatment will aim to emphasize that:

- the interventions aim to increase or regulate the fluxes of solar radiation and water through the crops;
- several important physiological processes are involved that interact in a complex way;
- the resultant input–response curves should be dissected to show which factors are operating and which can be manipulated most efficiently.

7.1.1 Managing water

Unlike solar radiation, the supply of water to soil and the use of water by crops are not simultaneous. Dryland crops commonly grow on the residual store of soil water that is available before sowing, without subsequent additions through rainfall or irrigation. Therefore, as argued in earlier chapters, water is a more widespread limiting resource than solar radiation. The relation between yield and the use of water by crops therefore provides a general quantitative means of comparing agronomic practices and crop varieties. This is true in principle at least, since few agronomic studies standardize responses with respect to the fluxes of water or radiation.

In Chapter 3, the concept of a transpiration equivalent, Ω_w (g kPa kg^{-1}), was introduced to describe the efficiency with which water captured by different crop species is 'converted' into the products of photosynthesis. This equivalent is a reasonably conservative trait within a crop variety, leaving scope for management primarily through the amount of water taken up and transpired by the crop. Roots must either be induced to grow towards the source of water or water must be re-directed to where the root system is or is expected to be during the season.

The amount of water that can be *captured* by plant roots can be regulated either directly through irrigation or through water harvesting methods that redirect a greater proportion of the annual rainfall to plant roots, or indirectly by applying nutrients that enhance root growth or by reducing competition from weeds. It is important to link any practical methods with the principles introduced in Chapter 3. In particular, the hydraulic release characteristics of different soils and the physics of evaporation and drainage need to be understood before any rational assessment can be made of water harvesting and irrigation techniques. Water management in cropping systems therefore has three objectives, to maximize or otherwise regulate: (i) the total amount of water available in the soil to plant roots; (ii) the fraction of available soil water that is transpired by the crop; (iii) the efficiency with which transpired water is converted into crop dry matter.

7.1.2 Nutrients

Soil fertility and its management have generated considerable study and have received their own detailed and comprehensive treatments to which the reader is referred (e.g. Wild, 1988; Syers, 1998; Giller, 2001). The yield of a crop obtained without fertilizer or other additives varies very greatly among agricultural regions and soil, and is a consequence of both the basal soil fertility and various climatic constraints. Basal fertility is reduced by continuous cropping, and was traditionally maintained by extended fallowing between crops. However, the rise of human populations in many agricultural regions now precludes fallowing. Taking extremes in a similar radiation environment, the basal yields vary by much more than 100-fold between that of a grain crop in some sub-Sahelian soils, where yield of reproductive parts can be constrained to 0.1 t ha^{-1} by a combination of low N and P in the soil and low rainfall, to that of perennial plantation crops on some coastal clays in South-east Asia, where rainfall and nutrients are hardly limiting at any time and yield can be as high as 20 t ha^{-1}. In severely infertile environments, yield can in principle be increased tenfold, in the short term at least, by addition of nutrients.

Plant nutrition is very complex for a number of reasons. First is the range of essential nutrients and the variety of their functions. Nitrogen (N), phosphorus (P) and potassium (K) are present in relatively high concentration in plant tissue, being *inter alia* constituents of proteins, lipids or osmotic agents, while trace elements such as iron and zinc occur in very low concentration, within certain enzymes for instance. Any interpretation of the input response needs an acceptance that the elements have different functions, and that they must interact through these functions. A second issue is that a soil's power to hold and release nutrients depends on the factors of soil structure and biodiversity that influence a soil's long-term cohesiveness and stability (Greenland, 1995; Syers, 1998). The plant's roots fuel the soil microbial populations, which in turn break down residues, take part in transforming minerals, and bind soil particles. Microbes also have major roles in mediating symbiotic nitrogen fixation by legumes (Giller, 2001) and, as *mycorrhiza*, in absorbing certain minerals, especially in perennials. Introducing and maintaining organic matter in soils is clearly necessary for microbial function, but organic matter is often in shortest supply where most needed. A third issue is that the fluxes of the main elements in the soil are part of global cycles that include transport of nutrients in surface and drainage water and losses to and deposition from the atmosphere. The management of soil fertility is therefore much more involved and involving than simply adding nutrients. Nevertheless, adding mineral fertilizer, especially phosphates, to infertile soil is commonly the quickest way to boost yield above the most meagre levels (Fussell *et al.*, 1987; Bationo *et al.*, 1989).

Local solutions to specific problems of nutrition are often found through experiments on the type and form of the fertilizer, its timing and placement, the different release rates of formulations, especially of phosphate, the contribution of organic matter, and effects carried over from one season to another. Too often, however, work on nutrition is treated in isolation from other constraints and without regard to the broad mechanisms within the plant that are affected. A greater generality and transferability of results would be the case if nutrients were considered to operate by regulating the fluxes of radiation and (particularly) water. Examples of a modern methodology, including techniques for measuring nitrogen in plants and soils, are given by Haverkoort and Mackerron (2000).

7.1.3 Weeds

Weeds respond to environment in much the same way as crop plants. By competing for resources, they increase the stand density that each individual in the crop experiences.

They sometimes alter the epidemiology of pest attack by providing refuges for pest organisms and their predators. They are represented in many different plant families and in a range of life forms, but their study is to some extent aided by the consistent approach that weed biologists have developed and by the fact that important weeds are found in many parts of the world. Cousens and Mortimer (1995) provide a framework for studying weed populations, Evans (1993) considers the broader role of weeds and herbicides in the evolution of crop yield.

Weeds exert their effects in several ways. Those overtopping the crop canopy will reduce radiation capture by the crop by direct shading but might increase the conversion efficiency as in some cases of intercropping. Weeds forming a ground cover layer will hardly reduce capture by shading but could reduce conversion efficiency if they sufficiently deprive the crop of nutrients; capture might then decrease through reductions of leaf and root surfaces. Additionally, parasitic weeds, such as *Striga* spp., tap directly into the carbon source within the plant, and also are unlikely to reduce capture directly, but through carbon-limited surface expansion. Certain plants also interact by chemicals, either emitted while they are alive or from their decaying tissues when dead. Both are commonly defined by the term 'allelopathy', but some authors prefer the stricter use of this term to denote the emission and effect of chemicals while the plant is alive. Less well known and understood are interactions through positional sensing of one plant by another, for instance, through a plant changing its size in response to the balance of wavelengths around it. Many experiments on crops and weeds do not attempt to distinguish between these types of interaction, though in most circumstances, the dominant form of interaction is through direct competition for resource.

With reference to Fig. 7.1, the base yield in the absence of management connects in a complex manner with the corresponding base for nutrients. Unchecked weed populations are likely to be higher on fertile soils, and vice versa. The base can range from zero, where weeds so dominate the crop that they prevent it from seeding, to virtually the plateau itself if previous weed control has reduced the populations to very low numbers. Even with effective chemical herbicides, control is expensive or time consuming, so much research effort is spent in working out the most economical means to take yield from the base to, or near, the plateau.

One systematic approach to crop–weed interactions in the field uses the relation between weed 'density' – the number of weed individuals per unit area of field – and percentage loss of crop (Cousens, 1985; Hughes, 1996). For consistency with the general input–yield approach in Fig. 7.1, however, we prefer to relate inputs (weed management) to yield *per se* between a maximum weed infestation that gives the base yield for a system and a minimum that is too small to reduce yield from the plateau. Whatever system is used to represent the effects of weeds, management regimes range between the use of thresholds, where control is applied only if certain weeds reach specified densities at specified times, and 'prophylaxis' where some form of control is applied repeatedly over periods of years in order to keep weeds below a specified minimum. Working to a threshold weed density might save time and effort in years when weed populations are poorly developed, but could require greater intervention (more time, more chemicals) if weed populations have already accumulated to well above the threshold value at the time of the inventory.

7.2 Empirical input–yield responses

The input–yield curve and associated statistical analysis are a system-scale description, similar to the relation between population density and yield. They define the limits in which the fine-scale physiological processes have operated in the particular system being examined. However, the curves tend not to be transferable between environments because they are made up of these complex interacting processes. We begin by exploring some typical responses illustrated by a range of case studies. Mead *et al.* (1993) provide statistical methodology.

7.2.1 Limitation by multiple factors

When studying any particular limitation, it is important to remove or reduce the effect of other possible limitations. In nutrient trials, weeds can be controlled completely, and land shaping employed to conserve as much water as is feasible. Multifactorial trials involving several nutrients should then demonstrate responses to each nutrient when others are non-limiting.

The interdependence of nutrients is shown by yield responses of pearl millet in the impoverished soils of the Sahelian region, cited by Fussell *et al.* (1987). The data, represented in Fig. 7.2, show a basal yield (without added nutrients) of about 200 kg ha^{-1}, a near linear rise in yield with added N when P and K are also added at non-limiting concentrations, and the range of yield with variation in P at moderate addition of N. Low availability of P is strongly limiting

yield in these soils and climate, as shown by the yield when P = 0 being hardly above the yield when all three nutrients are withheld, and below the yield when N=0 and the others are at moderate applications.

The maximum response in this experiment took yield to 1400 kg ha^{-1}, eight times greater than the basal yield. The shape of the curve suggests that the system still lay on the slope AB in Fig. 7.1, and further increases were possible.

7.2.2 Temporal variability and trends

Variability in the weather, particularly the rainfall, between years can obscure responses to cultural factors (metaphorically speaking, the noise being high compared to the strength of the signal). Means should be explored to reduce the 'noise' and reveal the underlying trend or 'signal'.

An example of order arising out of variability is given by experiments on the effect of tied furrowing in crops of maize, sorghum and cotton grown in Zimbabwe, as reported by Jones *et al.* (1989) and reinterpreted by Squire (1990). The soil was at least 1 m deep and highly fertile, so that nutrients probably

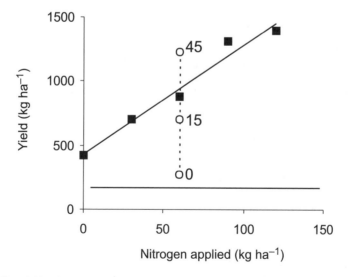

Fig. 7.2. Pearl millet yield response to nutrients in an impoverished Sahelian soil, illustrating the basal yield (A in Fig. 7.1), the response to added nitrogen (closed squares) and the response at 60 kg ha^{-1} nitrogen to additions of phosphorus between 0 and 45 kg ha^{-1} (open circles). The regression is $y = 8.5x + 434$, $r^2 = 0.97$. (Original data presented by Fussell *et al.*, 1987.)

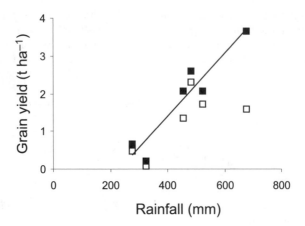

Fig. 7.3. Effect of water conservation practice on the response of yield to rainfall over 7 years in the lowveld of Zimbabwe: closed squares, tied furrows; open squares, unshaped soil. The regression is $y = 0.0083x - 1.93$, $r^2 = 0.9$. (Original data in Jones *et al.*, 1989.)

did not limit yield during the period. The basal yield, without furrowing, varied greatly between years but not just in relation to the amount of rain (Fig. 7.3). Years of few, heavy rainstorms gave little yield because much of the water was lost as runoff. Effort expended in 'tying' the furrows to contain rain water, encourage percolation and prevent runoff, caused yield to be greater in several years and more closely related to the amount of rainfall.

In data such as these, each pair of coordinates from any one year (the yield with and without tied furrows) lie on their unique input–yield curve, where the input is tying the furrow. The basal yield, without tied furrows, is variable between years because the characteristics of the rainfall differed between years. However, the set of maxima for each year appear to form a more general super-curve, for which the input becomes rainfall *per se*, not land shaping. One limitation (lack of water conservation) was overcome, and the stronger underlying limitation of rainfall was revealed. The linear relation between productivity and rainfall can be considered a strong signal emerging from the noise. The slope for the treatments with tied furrows gave no indication of levelling off, which implies that yield was still on the part AB of the graph in Fig. 7.1. (The

factors underlying the slope and intercept are considered later.)

The response can be more difficult, but still possible, to reveal when soil fertility is declining as well as the rainfall being variable. Information on yield and nitrogen recorded by Nader and Faught (1984), reinterpreted by Keating *et al.* (1991), demonstrate several general effects. Maize was grown on seven successive occasions during 1979/82 in the two rainy seasons at Katumani in dryland Kenya, during which rainfall varied between 200 and 500 mm a season, the former severely limiting yield, the latter allowing a maximum yield of around 4 t ha^{-1}. The results are plotted here as the yield at zero nitrogen against the yield at 50 kg ha^{-1} of nitrogen, half the total application (Fig. 7.4). The plot shows: (i) the basal yield in the wetter seasons declined over time as the level of nitrogen in the soil decreased, (ii) yield in wet seasons became more responsive to nitrogen as the basal yield decreased, indicated by greater divergence from the 1 : 1 line, and (iii) yield in three seasons of low rainfall was much lower than in wetter years and not greatly responsive to nitrogen. The inset in the figure shows the corresponding plot of yields at 50 and 100 kg ha^{-1}. Here the two yields still show the range related to rainfall but are

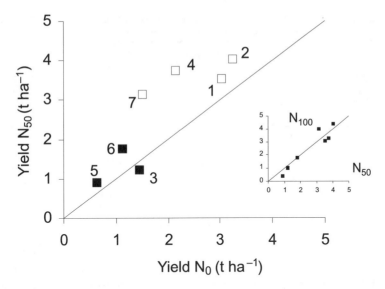

Fig. 7.4. Yield without N fertilizer plotted against yield with 50 kg ha^{-1} N over consecutive cropping seasons (numbered 1 to 7): open squares, high rainfall; closed squares, low rainfall. Inset shows the corresponding relation for yields at 50 and 100 kg ha^{-1} N. (Original data in Nader and Faught, 1984.)

much closer to 1 : 1, there being little further change in yield between half and full application. There was also evidence in the data that nitrogen unused in one dry season was not depleted from the soil profile but used by the plants in the following season.

The main agronomic uncertainty at this site was in the initial rise in yield with added nitrogen. The basal yield was not a good predictor of yield at 50 kg ha^{-1} of nitrogen, because of the decline in soil nitrogen and the suppressive effects of rainfall. There was an indication in the seventh season (see the original paper) that if the basal nitrogen continued to decline substantively yield would become responsive over a wider range of applied nitrogen, such that yield at 50 kg ha^{-1} of nitrogen would then not be a good predictor of yield at 100 kg ha^{-1}.

7.2.3 The contribution of genotype

The form of the input response is also sensitive to the nature of the crop genotype, especially its ability to forage for and compete for resource. Among a typical range of advanced genotypes in any crop species, the variation in yield at optimal input is usually less than the variation under limiting conditions. The effects of genotype on the input–yield curve are now shown for two case studies, which show similar responses despite one being on rice in West Africa and the other on cassava in Colombia.

Fofana *et al.* (1995) grew 12 varieties of paddy rice (*Oryza sativa*) and upland rice (*O. glaberrima*). Nutrients and water were not strongly limiting; the variable input was control of weeds, mainly *Brachieria lata*, *Trianthema portulacastrum*, and *Ageratum conyzoides*, at two levels – full control (BC on Fig. 7.1) and no control (the base level A). The results illustrate a number of the principles underlying weed–crop interactions as well as the general effects of genotype on the input–yield response. First, the variation in yield among varieties was much less under maximum input (full control) than at the base level without control (Table 7.1). Moreover, the weeds and crop together produced more biomass than the weed-free crop alone, implying that the weeds intercepted and converted additional resources than the amount they 'took' from the crop.

Table 7.1. Mass of crop yield and weeds (kg ha^{-1}) in a trial in West Africa comparing 12 varieties of rice grown with and without weed control. (From data in Fofana *et al.*, 1995.)

	Crop yield (full control)	Crop yield (no control)	Weed biomass (no control)	Weed mass + crop yield (no control)
Mean of 12 varieties	2607	1059	2329	3387
Standard deviation (SD)	426	477	743	403
SD/mean	0.16	0.45	0.32	0.12

Since also the variation between cultivars was least for the crop and weeds combined, the weeds were compensating for the yield of poorly competitive cultivars, so as to bring the combined biomass nearer to a biomass ceiling for the site and year.

In the treatments without weed control, weed mass and rice yield were strongly inversely related. The competitiveness of the varieties probably drove the relation; that is, the poorly competitive varieties allowed a high weed biomass that further restricted yield. The information can be summarized as something analogous to an input–yield response if the input (x-axis) represents competitive weed suppression as a percentage, ranging from 0% in the absence of a crop variety (no suppression) to 100% in the presence of a variety that suppressed weeds totally. No weed-only treatments were introduced in this case, so 0% represents the treatment (= variety) that allowed the maximum weed mass (3700 kg ha^{-1}). The maximum suppression of weeds by a variety was 74%, a reduction from 3700 to just below 1000 kg ha^{-1}. The coordinates form a linear trend which approaches the average weed-free yield of 2600 kg ha^{-1} at 100% (Fig. 7.5). The value of expressing results in this way (rather than a yield loss curve, for example) is that it treats genotypic variation as an input, thereby allowing direct comparisons, primarily, with control of weed mass by agronomic means and more broadly with other forms of input such as nutrients.

The final case in this section is taken from an experiment on cassava in Colombia (El Sharkawy and Cadavid, 2000) in which 14 varieties were grown in each of 5 years at 0, 50, 100 and 200 kg ha^{-1} of added potassium (K) (Table 7.2). The effects of K were

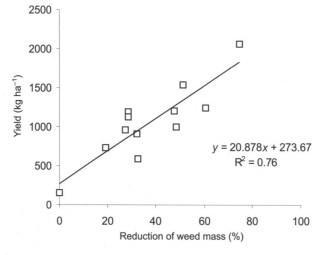

Fig. 7.5. Increase in yield caused by competitive reduction of weed mass (as a percentage of maximum weed mass in unweeded treatments) among 12 varieties of rice, *Oryza sativa* and *O. glaberrima*, in West Africa. The linear regression is $y = 20.8x + 274$, $r^2 = 0.76$. (Original data in Fofana *et al.*, 1995.)

Table 7.2. Dry root yield of cassava at four amounts of applied K fertilizer, showing mean yield (kg ha^{-1}) of the 14 varieties and the associated variance and range. (Original data in El Sharkawy and Cadavid 2000.)

Applied K (kg ha^{-1})	0	50	100	200
Year 1:				
Mean	13.6	15.9	15.5	15.2
Variance	2.67	2.79	2.25	3.94
Range	6.1	5.8	6	7.9
Year 5:				
Mean	5.31	8.49	9.81	9.71
Variance	2.66	3.22	6.96	7.23
Range	5.1	5.9	8.5	10.3

small in the first year when soil reserves were still high, yield being maximum at 50 kg ha^{-1}, rising no further at higher additions of K. The variance was smallest (but only just) at 100 kg ha^{-1} but increased at the next increment because this high level of K caused a decrease in yield of some varieties. After 5 years, a decline in soil fertility had caused yield without added K to be less than half what it was in the first year. Confirming the generality stated earlier, the variance and range were very much greater under K shortage than in the first year, especially in relation to the mean, showing that the varieties differed greatly in their response to low levels of K in the soil. The maximum yield was reached at 100 rather than 50 kg ha^{-1}, but again the variance and range continued to increase to the highest application rate. Within this variation, the authors were able to identify varieties that had a high base yield under potassium shortage and that also responded well to additions of potassium.

In concluding this section, we suggest that the large potential effects of agronomic intervention must be quantitatively measured through the same currency against the potential advantages brought by plant improvement. In the short term at least, and in impoverished soils, very large increases in yield, often of an order of magnitude, can usually be made through agronomic intervention alone. Substantive understanding of a system can be gained by empirical means in several years, though probably 5–10 years

of experimentation at the same site are needed to distinguish trends such as declining soil fertility from annual variation in the weather.

7.3 Understanding the input–yield response

As indicated, the input–yield response at any site is an empirical guide to practice rather than a definitive biological characteristic. If its form and magnitude can be explained, however, it can be used with greater confidence as an interpretive tool and extended to more surely explore new circumstances.

7.3.1 Global and local limitations

In the simple scheme in Fig. 7.1, crop stands function between a basal (A) and a maximum yield (BC). In reality, the position of the crop stand with respect to these limits is rarely fixed within a growing season, but changes as resources are used or replenished. Two measurable growth curves can therefore be defined: an upper one in which water and nutrients are plentiful, and competition with weeds is absent, such that the rate of growth (of a genotype) is limited by global factors such as solar radiation and temperature; and a lower one in which resources are not augmented and the rate of growth depends on local factors, including the capacity of the stand to forage for radiation, water or nutrients by extending its root system or canopy, usually in competition with weeds. The latter state, or something near it, is the commoner in most forms of tropical agriculture, where for instance a stand is extending its roots to access and deplete a store of water in the soil. The actual value of growth in this state depends on factors such as the rate of extension of plant surface, the rate of uptake of scarce resource and the transpiration equivalent.

The states are analogous to the exploitative and explorative functions of biological structures generally. Crops will move towards one state or the other during

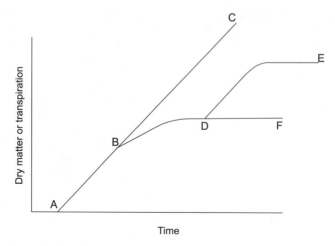

Time

Fig. 7.6. Dry matter production when there is no shortage of resource (ABC) gives way to production under a foraging strategy (BDF), until resource applied at D causes the rate to return parallel to BC until the resource is again depleted.

growth, depending on local conditions and management. For illustration, a representative growth curve under global limitation is shown in Fig. 7.6 as the line ABC. The diagram is adapted from Monteith (1986b) who offered it as a scheme to interpret crops' responses to irrigation. It can be applied to any input, however. By way of example, a stand following ABC becomes limited by water or nutrients at around point B, and begins to follow the slope of the curve BDF; that is, it adopts a foraging strategy. At point D, the limitation is relieved and the crop returns to grow along a path that is parallel to ABC, at least until the added resource is used, when the slope falls back to that of DF. Such changes occur at and between irrigations or applications of nutrients or weed control.

It is of benefit to appreciate where a stand is in such a scheme at any time. The curve ABC in Fig. 7.6 can be calculated from incoming solar radiation, the maximum seasonal curve of fractional interception, and an appropriate conversion coefficient, ε_s (for example, see Chapter 2 in this text and Chapter 3 in Squire, 1990). This curve provides a mark against which to compare the current performance of the crop, and assess the likely benefits of inputs, including weed control and fertilization. It also has more

practical benefit as a means of estimating the amount of irrigation water needed, by use of the transpiration equivalent (see later).

Global limitation and assessing irrigation needs

There are many methods of estimating the irrigation requirements of crops. Most of those in current practical usage descend from Penman's equations on evaporation. They rely on the use of meteorological data to estimate the potential evaporation rate from an ideal cover of vegetation (e.g. one completely covering the ground and photosynthesizing at maximum rate). Accounting for the other main components of the hydrological cycle then leads to an estimate of how far the available water falls short of potential evaporation. The shortfall can be taken to be the irrigation requirement. A combination of theory and experience rapidly led to standard methods of estimating water requirements, whereby potential evaporation was multiplied by crop coefficients which account for surface characteristics of the crop and include the reduced requirement while the canopy is forming (Doorenbos and Pruitt, 1977).

In terms of the scheme used in this chapter, irrigation should take the crop from the base to the ceiling level, or from curve ABDF to ABDE in Fig. 7.6. This offers the opportunity to estimate the demand for water by expressing it through the maximum rate that the crop can produce dry matter in that environment, i.e. the curve ABC in Fig. 7.6, and equating dry matter with an amount of water using the transpiration equivalent (Azam-Ali *et al.*, 1994). The curve of dry matter production can be estimated from values of solar radiation, together with knowledge of the seasonal curve of fractional interception (*f*) and the conversion coefficient, either culled from the literature or measured locally. A useful tip is that the logistic-type curve, commonly describing the seasonal course of fractional interception, is of such a form that about 35% of the incoming solar radiation is intercepted up to the point where the canopy achieves maximal *f*, say just before flowering in a short season cereal. Mean *f* over the life of the crop can therefore be estimated from the durations before and after that point.

These methods of estimating water requirements rely on prior knowledge either at the site in question or at comparable sites. The relevant input data might not be available to the tropical agronomist. Other, more practical, methods can still be used, however.

Some practical examples of manipulating water flux

An alternative to the physical and physiological methods described above is to use visible signals, such as the physiological condition of the crop, to indicate when irrigation is required. Plants must continually balance the demand for transpirational loss (in exchange for CO_2 gain) with the supply of water available to their roots. At the same time, plant water status must be regulated to maintain essential metabolic functions and plant structure. When the supply of water to and through plant roots is unable to meet the transpirational demand, plant water status is reduced and the plant is considered to be under water stress. Although a number of physiological (e.g. increased stomatal resistance) changes occur in response to this stress,

the first visible indicator that a crop is under water stress is often leaf wilting. In principle, visible wilting can be used to establish the need for irrigation in a wide range of crop plants. However, the problem is that by the time a plant shows symptoms of wilting it has already been under stress for some time and has therefore lost potential growth because of the reduction in CO_2 uptake as a consequence of stomatal closure. In these circumstances, regardless of the frequency and amount of irrigation applied, each wilting episode (allowing the crop to drop to line BDF in Fig. 7.6) contributes to a loss in potential yield. Ideally, irrigation should be applied before individual plants experience stress but not in such large amounts or frequencies that allows water to drain beyond the root zone. In theory, accurate measurements of soil moisture, using instruments such as a neutron probe, or physiological signals, such as changes in stomatal resistance or plant water status, could provide a basis on which to calculate irrigation need. However, such methods are unrealistic in terms of the practical management of irrigation systems at the field scale in all but the wealthiest regions of the tropics.

An effective alternative is to devise agronomic strategies that indicate when water stress is imminent for any particular crop so that irrigation can be applied before symptoms are evident. Figure 7.7 provides examples where visible evidence of wilting (but not of the target crop) can be used as an appropriate signal that irrigation is required. In each case, the particular agronomic intervention elegantly demonstrates how scientific principles can be applied in practice. For example, in Fig. 7.7a, a number of pits are dug within a field and the soil carefully excavated to approximately the rooting depth of the crop. This soil is then mixed with a quantity (say 5% by volume) of sand and then replaced within the original pit. In this way, a number of inspection positions are created within which the soil has a lower waterholding capacity than the surrounding crop area. The crop is then sown across the whole area in the usual way after suitable markers have been positioned above the inspection pits. By carefully observing the growth of crops within these inspection

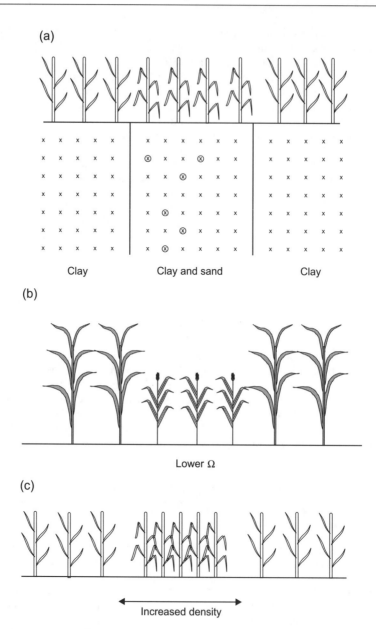

Fig. 7.7. Examples of how visible plant stress, i.e. wilting, can be used to identify irrigation need. (a) A number of pits are dug within a field and the soil is then mixed with a quantity of sand. The areas of soil with a lower waterholding capacity show symptoms of water stress earlier than elsewhere in the same field. (b) Differences in transpiration equivalent between species can be used to indicate the need for irrigation by inserting indicator plants with a lower transpiration equivalent within the main crop. (c) The indicator crop is sown at a much higher planting density than the main crop at specified locations where the plants show signs of wilting earlier than the main crop.

areas, the grower is able to monitor the onset of wilting at these locations. The principles of soil water dynamics introduced in Chapter 3 show that under otherwise similar conditions the areas of soil with a lower waterholding capacity will show symptoms of water stress earlier than elsewhere in the same field. By implication, the water stress that is evident for plants within the inspection areas is imminent for the remainder of the crop and it is therefore time to irrigate.

In Fig. 7.7b, the same principle can be applied without the need for soil excavation. Here, at a number of designated locations the crop is sown at a much higher planting density (say, 400%) than the main crop. Since plants within these inspection locations will be competing more intensively for the most limiting resource, in this case water, they will exhibit wilting symptoms earlier than the surrounding vegetation. Again, careful observation of the inspection areas for signs of wilting will indicate that irrigation is required. In Fig. 7.7c, the concept of differences in transpiration equivalent between species can be used to indicate the need for irrigation. Here, indicator plants with a lower transpiration equivalent are sown within rows of the crop to be irrigated. Since they are likely to suffer from

water stress earlier than the main crop, evidence of wilting in these plants can again be used as a signal for irrigation.

Clearly, all the examples above still require fine tuning, experience and testing by the grower to establish the best combination of species and method. Also, none of these methods provides an estimate of the irrigation need in quantitative terms. This remains a difficult area for simple agronomic interventions. Nevertheless, the methods illustrate how simple techniques can be used to complement and often replace more elaborate methods that may be unavailable or inappropriate to the small-scale grower.

7.3.2 Interpretation of the response surface

The empirical response to input in any crop in effect represents the transition between the curves ABC and ABDF in Fig. 7.6, integrated between germination and harvest. The nature of the transition is determined by factors of the spatial and temporal location of resource in relation to the plant and by factors of the plant's physiology. The basis of the input–yield curves for response to added nutrients is now demonstrated for a root crop and a cereal.

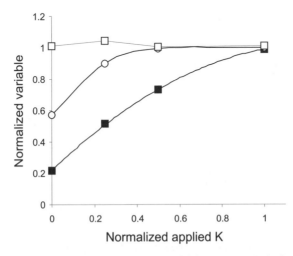

Fig. 7.8. Relation in cassava between applied potassium (K) (maximum 200 kg ha⁻¹) and K taken up by the plant (filled squares), total plant dry matter (open circles) and the fraction tuber/whole plant (open squares); all values normalized as a fraction of the highest. (Original data in El Sharkawy and Cadavid, 2000.)

The responses of cassava (El Sharkawy and Cadavid, 2000), referred to earlier (Table 7.2), are reproduced in Fig. 7.8, normalized as a fraction of the highest value in each case, where, for instance, a value of 1 for applied K corresponds to the full application of 200 kg ha^{-1}. Some general features of such responses include:

- the increasing amount of nutrient taken up at higher application but at a diminishing 'return', uptake being greater than applied at low application and less at high application;
- uptake of small amounts of nutrient at zero and low applications producing relatively large amounts of plant mass; and
- the curve for plant mass reaching a plateau or 'saturating' at moderate applications, while further uptake continued at higher applications. In this instance, the harvest index – the root yield as a fraction of the total dry matter – hardly varied over the range of plant mass generated, such that the curve for root yield against applied N was similar to that for total mass.

A more complex set of interactions defines the input–yield curve when low nutrient availability affects the allocation of N and dry mass among plant parts. The curve for response of maize yield to applied N is shown as the solid line in Fig. 7.9a

(Muchow, 1988a, b). When both variables were normalized as fractions of the respective maxima, which were 420 kg ha^{-1} N and 8600 kg ha^{-1} grain yield, the yield reached a plateau at around 0.6 of the highest value of applied N. The basis of the curve for yield itself can be seen in the corresponding curves for other variables. The N taken up, the total plant mass and the harvest index (= grain/total mass) began respectively higher and reached the corresponding maximum for the variable at lower levels of applied N. In cereals, the nutrient content of grain can be just as important as the mass of grain. Figure 7.9b gives the same type of graphical plot, but with uptake N as the x-axis, and with N in grain and nitrogen harvest index plotted. The curve for N in grain is strongly linear, while that for N harvest index starts high and reaches a plateau at about 0.5 the maximum applied N.

The responses in Figs 7.8 and 7.9 have a complex physiological basis. The diminishing return of the uptake curves lies variously in that uptake at high applications may be limited by growth rate (rather than the reverse), that root systems are unable to exploit the resource sufficiently and that the nutrients are lost from the vicinity of the systems. A small amount of N taken up at low applications causes a disproportionate increase of dry matter production which gets less and less as the physiological mech-

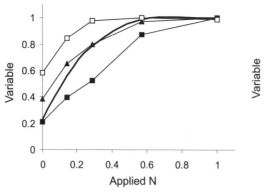
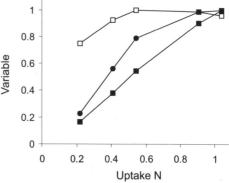

Fig. 7.9. Relations in maize between (a) applied nitrogen (N) and N taken up by the plant (filled squares), grain yield (continuous line), whole plant dry matter (filled triangles), and the fraction grain/whole plant mass (open squares); (b) N taken up by the plant and grain yield (filled circles), total N in grain (filled squares) and the fraction of N in grain over N in the whole plant (open squares). All values normalized as a fraction of the respective highest. (Original data in Muchow, 1988a, b.)

anisms saturate. The curvature (the diminishing return) in the relation between nutrient and crop dry matter is an amalgam of various effects on the interception and conversion of radiation. Generally, as in the work described by Muchow, the interception of radiation is readily stimulated at low applications through rapid enhancement of leaf area, then tends to a maximum at moderate levels of applied nutrient, largely because of the asymptotic relation between fractional interception and leaf area index (Chapter 2). Of the total solar radiation received in this example, about 30% was intercepted at the lowest levels of N uptake; the percentage rose, reaching a maximum of 60% at half the maximum N uptake. The effects of added nutrients would generally be much less than they are, however, if the stimulus of stand dry matter depended only on greater resource capture. In fact, the enhancement of the conversion efficiency (ε_s) continues at high applications, a result of the increasing N concentration driving a rise in the capacity of the photosynthetic apparatus to assimilate carbon (Lemcoff and Loomis, 1986).

Both of the examples just considered show that the fractions of dry matter and nutrients allocated to the harvestable parts are much less sensitive to applications than are most other variables. In the vegetative cassava tubers, there was no change, while the partition fraction (harvest index) for maize grain was reduced only at zero N and at the lowest of the four applications. Further analysis of the results reveals that low N reduced the allocation of dry matter to grain because the photosynthetic system collapsed during grain filling as N was moved out of leaves to grain. Some limit must therefore have existed on the mass of carbon compounds that could be retranslocated from other organs. Given these interactions, the strong linear relation between N applied and N in grain in Fig. 7.9b is particularly interesting. It is one of the few responses that is not asymptotic or shows a diminishing return. The main reason for this is that grains continued to accumulate N at high values of N uptake, even above the N level in the plant as a whole that saturated

the capture and conversion of resource by the crop.

If nutrients are presented as a concentration or percentage (weight of nutrient divided by weight of plant), the values generally decline as the plant develops but differ between organs. Leaves active in photosynthesis and seeds both have high nutrient content. Since nutrients are often recycled from old to new foliage, they can continue to generate the intake of carbon for long periods. The general decline over time occurs because an increasing fraction of the plant is composed of tissue low in nutrients, such as carbohydrate stores. Plants with a longer duration therefore tend to develop a lower percentage mass of nutrients. Organs such as cereal grains have a high concentration of nutrients and will 'demand' it from other tissues. There will commonly be a trade-off therefore between nutrient quality of grain and continued photosynthesis. Examples of seasonal growth patterns in the ratio for different species are given by Squire (1990, Chapter 6).

7.3.3 'Efficiency' in the use of resources

The notion of a 'use efficiency' to define the effect of inputs, is employed widely, and so requires comment here. Water use efficiency and fertilizer use efficiency (as commonly used) define the return of yield from the amount of respective input, so in practice they are determined for any crop stand as the ratio of the values on vertical and horizontal axes in Fig. 7.1. As emphasized earlier, 'use efficiency' or mass/mass ratios are inherently different for water and nutrients. Both have components that are external to the plant, that determine the placement of the resource for instance, and in the case of nutrients their chemical form. Once in the plant, water has the properties of a flux, while nutrients can be recycled internally. Moreover, the ratio for water depends on the saturation deficit of the atmosphere (Chapter 3), and is therefore much smaller in dry atmospheres, independent of anything else.

The ratios are therefore neither constant

nor an indicative trait of a species or geno-type. If the base level A at zero input is above zero yield, then the ratio must decline with rise in input, the more rapidly if the response diminishes between A and B. If yield does not begin to be generated until some way along the input axis, as in Fig. 7.3, then the ratio does the opposite, increasing with further input once some yield is gen-erated, at least until the yield increments begin to diminish on approach to the yield plateau. For instance, total plant mass per unit plant nitrogen mass in dryland grain cereals varies typically in the range 10–60, while the ratio for the whole plant may be twice as much. Values for purely vegetative crops can be substantially higher. The ratio for use of rainfall, even in dryland sites sub-ject to careful husbandry, can range from 1 to 6 kg ha^{-1} grain per mm water as rain or irrigation.

The experiments on maize Muchow (1988a, b) and cassava (El Sharkawy and Cadavid, 2000), referred to earlier, can be further analysed to show some main features of the mass/mass ratios for nutrients. In Fig. 7.10, uptake by the plant is normalized on the x-axis as a fraction of maximum nutri-ent uptake, the actual ranges being 40–200 kg ha^{-1} of N in the maize experiment and 30–130 kg ha^{-1} of K in the cassava. Lowest

values in the zero application treatments in each case were (coincidentally) about 20% of the highest. The ratio of the mass of whole plant to mass of respective nutrient in the whole plant (i.e. biomass/N and biomass/K, respectively) are similarly normalized on the y-axis. Both mass/nutrient ratios fall with increasing nutrient uptake in a broad-ly similar manner to reach 0.6 and 0.4 of the maximum in the stands of highest nutrient uptake. The continued fall in normalized ratio over the whole range was because nutrients continued to be taken up even when the crops' production systems had maximized dry matter production. The ranges of both normalized uptake and nor-malized biomass/nutrient ratios shown in Fig. 7.10 may be typical in fertilizer trials on nutrient-poor soils that have not been fertil-ized for several years.

7.3.4 Competition and spatial interactions

The argument and examples in this chapter have not explicitly included spatial varia-tion in either the resource substrate or the distribution of crop or weeds and other pests. A formal treatment of heterogeneity at the field scale is still lacking, but some of the issues are raised by a brief look at com-

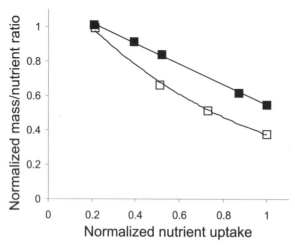

Fig. 7.10. Nutrient uptake in relation to the ratio of dry matter to respective nutrient in the whole plant, for nitrogen in maize (filled squares) and potassium in cassava (open squares); all data normalized as a fraction of the highest value. (Original data in Muchow, 1988a, b, and El Sharkawy and Cadavid, 2000.)

petition and aggregation in crops and weeds (Doyle, 1991; Hughes, 1996). Crops are spatially non-uniform because of sowing pattern (giving rise to degrees of rectangularity), variation in sowing frequency, emergence and predation, but weeds are more generally aggregated through preference for local microsites and through local dispersal patterns.

In practice, the potential of a weed canopy to divert resources away from a crop can be obtained by measuring resource use and conversion of crop and weed flora separately, and in different combinations of crop and weed. Such competition experiments comprise combinations of crop and weed in a replacement series (Connolly, 1986), in which the crop plants are increasingly substituted by weeds (density not variable) or weeds are increasingly introduced between the crop plants (density variable). Weeds of different competitive ability will give rise to different slopes and base values (as in Fig. 7.1), which give weed specialists information of the ranking of weed species as yield-reducers. Such differences between weed species can be substantial, 10- or 20-fold among important species, and allow targeting of weed control operations from early knowledge of weed population densities.

A more mechanistic approach has developed from de Wit's work on the physiological interactions between competing plants (Kropff and van Laar, 1993). Increasing infestation by weeds could result in reduced capture by the crop, reduced conversion efficiency or both, depending on the position and architecture of the weeds. Recent three-dimensional models of gas exchange, light and leaf position in weed/crop mixtures have given accurate simulations of the outcome of simple competitive interactions. While other stresses are not limiting, the partition of resource between the two plant types is largely determined by shading and gas exchange characteristics. Though it is difficult to measure interception by weed and crop plants *in situ*, models based on radiation physics in canopies confirm that crop biomass is reduced by weeds mainly according to the relative size and orientation of the foliage.

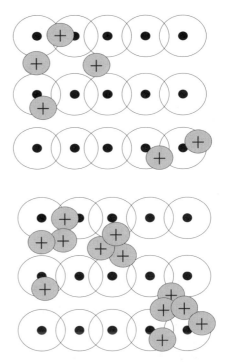

Fig. 7.11. Representation of crop plants (dots and open circles) and weeds (crosses and shaded circles) to show greater interference between weeds at high (lower three rows) compared to low density (upper three rows).

Experimental confirmation of the effects of aggregation has been slow to emerge, at least partly because of the difficulty of introducing aggregation patterns in a reproducible and systematic manner. However, there are grounds for supposing that aggregation in crops and weeds will affect yield in different directions. Wade *et al.* (1991) demonstrate by modelling how aggregation of the crop plants (without weeds) could reduce yield to 75% of that in a completely uniform stand, as the increasing aggregation creates both greater competition between a proportion of the individuals and more space that is not used. Weeds, if aggregated, achieve a similar effect by creating 'holes' of low yield in the crop canopy where weeds are most dense, but areas of less or negligible interference where they are present at low density. For any global density, however, yield should increase as aggregation, the degree of clumping, of weeds increases (Brain and Cousens, 1990; Hughes, 1996). The effects can be envisaged by the interactions in the diagram that is Fig. 7.11, where each plant is represented by a potential circle of influence on the soil surface (which would be a cone or cylinder in three dimensions) from which they tap resources. The weeds are shown aggregated at two densities; at the higher one they compete with each other as well as with the crop. Aggregation of weeds can be quantified by a range of statistical indices some of which are simple to calculate from data on counts of individuals in quadrats (Hughes, 1996).

The uncertainties in weed biology stem not so much from lack of knowledge of the mechanisms underlying their interaction with the crop as with their life cycle behaviour, persistence and spatial–temporal patterns. Agronomy could benefit greatly from adopting more widely some of the methods of plant population biology, which allow a practical understanding of the gains and losses in a weed population over time (Sagar and Mortimer, 1976; Cousens and Mortimer, 1995). Constructing a life cycle diagram readily shows that most weed species are able to produce very large numbers of seed relative to the number of plants of the next generation actually growing in the crop. This discrepancy is partly because the losses in the transitions are very considerable and partly because the next population of plants is strongly limited by microsites for germination, emergence and establishment. Much of the non-chemical control of weeds is based on either reducing the number of such microsites or increasing them temporarily so that weeds germinate in large numbers which can then be killed by tillage. The seedbank, of viable buried seed, is important in the life cycle of most weeds. Its description and analysis will be increasingly valuable for quantifying and interpreting shifts in the whole weed community, specially if the seedbank is considered to have more than one function, for instance a reserve of plant diversity as well as constituting a weed problem. For instance, plant community analysis was used by Chikoye *et al.* (1997) in Nigeria to describe change in the seedbank over several years during crop-

ping and fallow, and by Smith and Webb (1993) to separate the effects on the weed flora of location, fallow and manuring.

7.4 Scale and integration: extending the principles

In principle, the frames of reference, used within experimental fields, can be extended to encompass farming regions and farming systems.

7.4.1 Use and interpretation of a simple empirical relation

Even a simple plot of yield on a major variable such as rainfall can provide a necessary starting point for more detailed dissection of the input response. Huda *et al.* (1985) collated data on yield and rainfall for both research stations and farms at 15 locations in India. The spread of responses is summarized in Fig. 7.12. The most responsive research station fields gave yield/rainfall ratios of 6 kg ha^{-1} mm^{-1}, while the least responsive were no less than half that. In contrast, the ratios for farms were considerably lower. Full details are unavailable, but the steepest slope probably represents fields from which little of the rain was lost either as runoff or deep drainage, and crops that were stimulated by adequate nutrients to form an effective canopy and root system and a balanced internal allocation of dry matter. The yield/rainfall ratio of 6 kg ha^{-1} mm^{-1} equates to 0.6 g kg^{-1} (yield to rainwater). If gross assumptions are made that loss of water from the soil surface is 30% of rainfall, that the crops then extract 80% of the rainwater remaining in the soil, and allocate 40% of their total dry matter to yield, the 0.6 g kg^{-1} (yield to rainfall) then approximates to a dry matter/transpired water ratio of 2.7 g kg^{-1}, a value which was shown in Chapter 3 of this volume and in Squire (1990, Chapter 4) to be typical of a semi-arid environment with a high saturation deficit. Breaking down the input–yield response in such a way can provide the link between plant traits, environment and yield

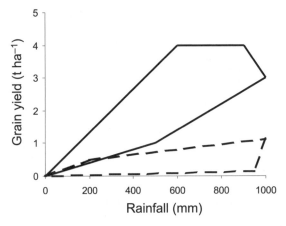

Fig. 7.12. Relation between grain yield and rainfall for cereals crops, maize, sorghum and pearl millet, at dryland sites in India. The solid line polygon shows the limits of the region occupied by sites on research stations receiving high inputs, the dashed line polygon sites on farmers' fields. (After Huda *et al.*, 1985.)

potential over a large geographical area. We enlarge on this in the following example of a national scheme to collate and synthesize input–yield responses.

7.4.2 Regional effects: the example of oil palm in Peninsular Malaysia

The coordinated work on oil palm (*Elaeis guineaensis*) in Peninsular Malaysia is one of a few examples of the way base conditions, local environment and fertilizer applied influence resource fluxes on a national scale. The trials on oil palm plantations were established between the late 1950s and early 1970s and run by several independent, private agencies. The aim was to find the most economical fertilizer regime for each of the various soil types on which the palms were grown. However, several developments, some in parallel initially, gave rise to a means of synthesis which resulted in adding considerable value to the basic set of measurements.

Four of the companies set up the Oil Palm Genetics Laboratory in the late 1960s, whose purpose was to study the basic genetics and physiology of the palm. Given the plant's great size, one of their first results was to derive stable stoichiometric relations between easily measurable morphological traits and the leaf area and the dry matter of plant parts (Corley *et al.*, 1971a, b). The crucial step for later physiological analysis was to have these measurements made routinely in most of the trials.

By the mid-1970s, the national agricultural research institute had collated and successfully analysed fertilizer response functions of some of the first wave of trials (Foster, 1976; Foster and Goh, 1977). They fitted polynomial functions to the input–yield responses to N, P, K and Mg fertilizer, were able to define non-limiting levels of each fertilizer in the various soils, and then fit the average response to each fertilizer at non-limiting levels of the others. In the early 1980s, the newly formed Palm Oil Research Institute of Malaysia (PORIM) continued to collate the agronomic and physiological information from a larger set of trials. By the mid to late 1980s, Foster and colleagues at PORIM had systematically fitted response functions to a wider range of soils and climates, and were able to identify other constraining factors such as low rainfall after application, low K buffering capacity and the effect of slope on reducing the response to nitrogen fertilizer. The result was one of the most complete studies of input–yield responses for any region.

As with other functions of this type, they had little basis in biological mechanisms, so had limited relevance to other issues such as the energy and water balances of plantation areas. It was also generally accepted that yield at sub-optimal fertilizer applications was limited by the element in question directly restricting oil production. However, further systematization was achieved by applying the principles of environmental physics and physiology (Squire, 1986). The first step was to show that the relation between leaf area index and fractional interception (*f*) of solar radiation (Chapter 2) was similar in a range of environments. The standard measures of leaf area, carried out in all the trials, could then be used to estimate fractional interception of radiation by each site and treatment. The total incoming solar radiation was obtained from the nearest office of the country's meteorological network and used to compute annual values of intercepted radiation by the stands. The conversion coefficient for solar radiation (ε_s) was calculated from the values of annual dry matter production, which were estimated from the morphological data collected for each treatment.

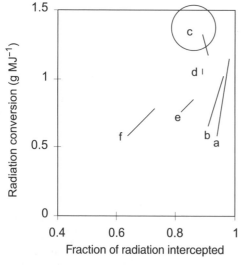

Fig. 7.13. Physiological response to fertilizer by oil palm in different soils of Peninsular Malaysia. Lines link paired values of fractional interception (*f*) and conversion efficiency (ε_s) in the treatments giving smallest and largest biomass at each trial. The circle indicates the region of maximum fruit bunch yield. Text gives explanation.

This procedure opened up three potential new uses of the data: the physiological limitation to yield (in terms of *f*, ε_s and partition fractions) could be defined for each trial; the physiological traits of oil palm could be compared with those of other world crops through the common currency of radiation flux; and the effects could be computed of base conditions and added fertilizer on the radiation, water and energy balances of large areas of the country. As an example, estimates of *f* and ε_s for trials on a range of soils are given in Fig. 7.13. The lines link the values in treatments giving the smallest and largest (plateau) yield at each of the sites (i.e. A to B in Fig. 7.1). The position of each line and the slopes provide insights on base nutrient conditions, initial physiological state, responsiveness to fertilizer and other limiting factors. Sites 'a' and 'b' in Fig. 7.13 are on inland soils that had been cropped for many years and had been reduced to a low nutrient content. These were older genotypes of oil palm that invested heavily in the vegetative, resource-capturing structures of the fronds and trunk. The *f*-values were large therefore, but ε_s values small, probably because low values of nutrients had reduced the efficiency of the photosynthetic apparatus. Fertilizer had little additional effect on *f* in these trials, but increased ε_s greatly (and thereby yield also). Site 'c' was on a coastal clay, high in organic matter, and consisted of a newer genotype which in this area had achieved the highest yield of oil palm in the 1980s. Its *f*-value was slightly lower than that of the older genotypes, a result of less investment in vegetative structures and more in fruit. Its base value for both traits was also high, a result of the high nutrient capacity of this soil. The fertilizer dose giving maximum yield actually reduced *f* (through smaller frond area) but increased ε_s higher values of biomass and yield were obtained, so the upper point on this line is the absolute upper limit for the genotypes and environments encompassed by the trial system. Site 'd' also had a high base fertility but was hardly influenced by added fertilizer: some other environmental or possibly a genotypic limitation was operating. Sites 'e' and 'f' were on 'problem' soils that had a low base fertility, such that the palms there had much less than optimal canopy area. Added nutrients caused increases in both traits about equally, but again there were severe environmental limits on the maxima that could be achieved. At site 'f', for instance, the limitation was most likely thin soil and low rainfall.

By expressing the effect of nutrients through the processes of interception and conversion, it was shown that yield was not so much controlled by direct effects on the biochemical pathways of oil synthesis, but by the interplay between frond area and conversion of radiation. The highest biomass was obtained when f and ε_s were both maximal, but because of the diminishing returns from investment in frond area, the highest yield occurred at an optimal f-value of around 0.85. To achieve a greater fractional interception would have caused diversion of large amounts of dry matter to vegetative matter and away from yield.

7.4.3 Farming systems and element cycles

This chapter has primarily considered the response to input and how this is composed of several interacting physical and physiological features. However, the input–yield responses at a site themselves influence, and are influenced by, the much broader issues of geochemical cycles and farming systems.

A more complete understanding of the role of agronomy in physical and social ecology in the tropics requires some understanding of the balance sheets for the main environmental fluxes and important elements. The earlier chapters gave background on the water and radiation balances. Comparable approaches can be undertaken for the gains and losses of nitrogen and for important minerals. In much of the tropics, the cycling of elements in the soil is very much influenced by animal husbandry, of which crop agronomy must become increasingly aware.

Similarly, agronomy must take its role

in the farming system through integrated management. In the case of weed control, for instance, 'integrated' means management that takes account of a range of temporal and spatial processes rather than simply killing a weed population at one time. Integration can be attempted and achieved at a range of scales including the weed patch, the field, the whole farm, groups of farms, the landscape and the ecological region. The smallest and largest scales are coupled through seed dispersal and gene flow: the arrangement of fields and the practices in them influence the regional connectivity for propagules; this connectivity itself feeds back into local practice by altering the range and physiology of weeds that affect the crops. If it is feasible to manage at larger scales, the individual holding may be less exposed to uncontrolled influxes of weeds. However, to integrate weed management at any of these scales requires knowledge of what drives both local dynamics and long distance spread of weeds. A practical example of using knowledge of weed biology is the use of catch cropping to control *Striga*, where a crop of, for example, Sudan grass (*Sorghum sudanense*) is grown to allow infestation by the parasite and is then uprooted thereby removing the parasite (Oswald *et al.*, 1997). Moreover, a survey of *Striga hermonthica* in Western Kenya illustrates the insights gained by regional as opposed to only a local approach (Frost, 1995).

Integration also implies carrying out fertilizing, water harvesting and weed control as part of a number of activities that together achieve aims such as greater production, more economical farming or reduced pollution. Employing a rotational sequence that varies the environment, such that nutrients are encouraged to accumulate and that no single type of weed dominates, is one form of contribution to integrated management. The rotation is likely to be a means of optimizing several processes, including weed numbers, soil fertility and the dual needs for fodder and food. Here, the aim is not to maximize any one of these aims but to find some skilful way to optimize them. The main point that we wish to reiterate is that progress as a whole in tropical agronomy will be surer if empirical responses are understood and compared in terms of their effects on the fluxes of solar radiation and water.

8

Synthesis

This chapter summarizes some of the historical developments responsible for the current state of global crop production. Options for the future are then considered, including broadening the base of crop types and species in tropical agriculture. A case study is given for the rapid development of knowledge in the underutilized tropical legume, bambara groundnut. The book ends with a discussion of the role of scientific agronomy as an intermediary between genetic technology and ecological principles.

8.1 Some major developments in agricultural research

Agriculturalists have experimented with different management practices and have selected the most promising plants from their own crops for over 10,000 years. There is, therefore, a long history of local agricultural research throughout the world. However, the formalization of agronomy and plant breeding into discrete academic disciplines and their contribution to the dramatic increases that have occurred in global crop yields are much more recent phenomena. The contribution that plant breeders, physiologists, agronomists, soil scientists and others have made to achieving global food security is at least as significant as any of the great political and social revolutions of the 19th and 20th centuries. Evans (1993) provides a compelling account of how historical increases in world population over the last century have been at least matched or surpassed by contemporary increases in the grain yields of the major food crops. For example, over this period, the yields of wheat in England and rice in Japan have increased from about 2 t ha^{-1} to over 6 t ha^{-1}, while world population has increased from less than 2×10^{12} to approaching 6×10^{12} people. These dramatic increases in food production, achieved without an overall increase in cultivated area, are largely due to the successes of agricultural scientists over the past two centuries and the integration of chemical, physical and biological principles within the armoury of agricultural science.

The usefulness of inorganic fertilizers was first demonstrated by Von Leibig (1803–1873) and Boussingault (1802–1887) who independently showed that the application of chemical products derived from nitrogen, phosphorus and potassium could dramatically increase the yields of crops beyond those already achievable through the use of manures (Marini-Bettolo, 1987). In the early 1900s, the large-scale production of fertilizers from atmospheric nitrogen was made possible through the industrial production of ammonia via the Haber Process. This allowed the commercialization of agriculture, at least in those industrialized countries where nitrogenous fertilizers could be easily and cheaply produced and distributed. At about the same time, contemporary developments in soil chem-

© CAB *International* 2002. *Principles of Tropical Agronomy*
S.N Azam-Ali and G.R. Squire

istry and plant physiology, and the genetic selection of potentially valuable cultivars (cultivated varieties), led to large increases in agricultural production. Until then, increases in the total area under cultivation had been the main source of increased food production. However, developments in agriculture now allowed crop yield per unit area of land to increase alongside the incorporation of hitherto marginal lands into agriculture through new agronomic practices and the breeding of robust varieties. For the most part, these advances were restricted to the Northern Hemisphere where economic development, the specialization of labour and the onset of mechanization all led to innovation and great changes in agricultural practice.

In the 1930s and 1940s, developments in industrial chemistry provided the basis for the commercial production of pesticides (DDT, chlorinated hydrocarbons), fungicides, nematicides and herbicides. Agriculturalists were now able to protect their potentially higher yielding crops against competitor plant species, predators and disease organisms. Again, the result was a significant increase in the yields of crops particularly in the developed regions of the world.

8.1.1 Global crop production and the Green Revolution

The rapid developments in agriculture in the first half of the 20th century were further accelerated in the 1950s and 1960s when production systems for some of the major world crops were transformed by what became known as the 'Green Revolution'. This was a technological package that combined potentially high yielding varieties of certain crops, principally wheat, rice and to a lesser extent maize, with inputs such as fertilizers, pest and disease control and irrigation. At the end of the Second World War, Dr Norman Borlaug and his colleagues achieved the first products of the Green Revolution through the breeding of so-called 'dwarf' varieties of wheat in Mexico. By 1956, Mexico was self-sufficient

in wheat and subsequently became a temporary net exporter of cereal grain. Experimentation with the dwarf Mexican varieties began in India and Pakistan in 1963 and by 1969, 55% of the 6 million hectares of wheat in Pakistan and 35% of the 14 million hectares of wheat in India were sown to Mexican varieties or their derivatives. In both India and Pakistan, the introduction of semi-dwarf varieties meant that wheat production approximately doubled between 1966 and 1971 (Hanson et al., 1982).

The advantages of potentially higher yielding varieties with better methods of growing and harvesting them meant that 'Green Revolution technology' was rapidly adopted by most farmers in the developed world. It was also embraced by those farmers in the developing world who had access to cash or credit facilities that allowed them to purchase the required technology. In essence, the Green Revolution was ideally suited to those regions where there had already been some degree of industrial development, in particular where there was cheap and readily available access to sources of fertilizers and irrigation and established methods of transporting and marketing products through regional and international trade. Although Dalrymple (1985) noted that semi-dwarf varieties of wheat and rice 'spread more widely, more quickly, than any other innovation in the history of agriculture in the developing countries', Gill (1991) demonstrated that the distribution of the most successful high-yielding varieties of wheat and rice was extremely uneven among different regions of the developing world. He further noted that 'in particular, sub-Saharan Africa, with its appalling record of famine, has reaped fairly insignificant benefits from these crops'.

Besides an increase in the yield per crop, two other factors have influenced the expansion of the global harvest. The first is that, where the annual cycle of temperature and either rainfall or irrigation is sufficient to support crop growth throughout the year, further intensification of agriculture is possible by increasing the frequency of cropping. This means that two or even three crops can now be grown on a piece of land that would otherwise have produced only one crop per year. To achieve this, modern Green Revolution varieties of cereals such as rice are photoperiod-insensitive. In circumstances where other conditions such as soil temperature and moisture content are not limiting, photoperiod-insensitive crops can be sown at any time of the year. Of course, this intensification brings with it the need for greater inputs of fertilizers, irrigation, pesticides and herbicides.

The second factor has been a concentration of effort into a few species and a consequent reduction in the number of species cultivated and the diversity of cropping systems. Many traditional food crops have been replaced by a small number of staple crops capable of achieving higher yields in response to fertilizers and irrigation across a wider range of environments. This has led to the further expansion of major crops, such as wheat, rice and maize cultivated as monocultures, into areas previously occupied by a diverse range of local crops and cropping systems. There are a number of texts on the consequences of the Green Revolution on tropical agriculture. Amongst these, the reader seeking further information should consult Evans (1993), Hanson *et al.* (1992), Gill, (1991) and Hazell and Ramasamy (1991).

8.1.2 The common cereal ideotype

A pre-requisite of the Green Revolution was the production of high yielding varieties (HYVs) of major staple crops. Strictly, we should refer to these as potentially HYVs since the yield increases that they can deliver depends to a large extent on the application of irrigation, fertilizers and other inputs. Nevertheless, the development of potentially higher yielding varieties has been the result of plant breeding programmes directed to achieving specific objectives.

Successful selection for visible traits, such as resistance to specific pests or diseases, is often easier than achieving more complex objectives such as grain yield or

greater efficiencies in the use of resources. With this background, plant breeders concentrated their efforts on increasing or stabilizing the yields of the major food crops in environments where new varieties were likely to achieve their yield potential. The ability to achieve this was eventually aided by a more systematic approach to plant improvement, incorporating the whole plant. Traditionally, much plant breeding had taken two approaches, 'defect elimination' and 'selection for yield'. Defect elimination includes the breeding of resistance or avoidance characteristics into susceptible but otherwise desirable genotypes. Here, the criteria for selection include the specific traits that correct the defect. An example is the incorporation of early maturity into varieties that may be vulnerable to water stress later in the season; other traits such as height of stem might change in consequence. 'Selection for yield' requires no direct attempt to incorporate specific physiological or morphological characteristics but, rather, attempts to increase yield irrespective of the means by which this increase is achieved. In some circumstances, selection for yield may simply involve hybridization between two parents defined as promising because they have individually shown good ability for yield. The success of such methods depends on the availability of a wide range of potential material within the breeding programme, the choice of crosses to be made, the evaluation of emergent genotypes, intuition and luck. The breeder may not know why a particular variety outyields its predecessors nor have any idea of the processes underlying the yield advantage. It is often left to physiologists to subsequently examine and explain the reasons for the increased yield.

A more systematic approach was formally introduced by Donald (1968) in his definitive paper where the concept of the crop *ideotype* was outlined – literally 'a form denoting an idea' – in which he argued for a rational basis to the breeding of model or designer plants. Despite initial and more recent scepticism (Balkema-Boomstra and Masterbroek, 1993) many of the ideas first proposed by Donald now form part of the standard inventory of plant breeding programmes. Donald argued that, as with the design of aircraft, buildings and instruments, the design and breeding of crop plants could usefully be centred on a theoretical model based on the knowledge, experience and imagination of agricultural scientists. He suggested that although our knowledge of the anatomy and physiology of some crop species may be insufficient to design new cultivars, in others, notably the cereals, a model of superior productivity might be feasible. He made the point that however crude the model might be as an approximation of reality, it provides the opportunity to devise and examine a combination of characters that might otherwise not occur in breeders' plots for centuries. Further, models provide new bases for the understanding of crop ecology and the design of more effective models. Crucially, Donald pointed out that 'selection for yield alone is unlikely ever to approach the asymptote of yield, since the appropriate

combination of plant characters, never being selected for, can be attained only by attrition or chance'.

The concept of a crop ideotype was first articulated for wheat production in temperate regions. Subsequently, the International Rice Research Institute (IRRI) based in the Philippines applied similar concepts to the development of an ideotype for rice, IR8, that was first released in 1966. Dingkuhn *et al.* (1991) used crop simulation modelling to extend the original concept of the rice ideotype. The ideotype concept and related activities on crops such as wheat and rice have important implications both for the range of crops and cropping systems in the tropics. Before exploring their implications, we first summarize the characteristic features of wheat and rice ideotypes.

The wheat ideotype

Figure 8.1 illustrates the architectural features of a wheat ideotype as proposed by

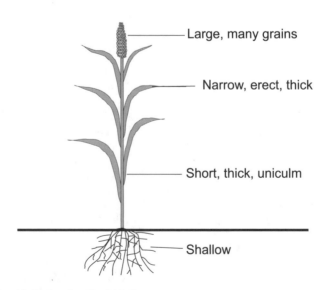

- Short duration
- Daylength insensitive
- Pest and disease resistance

— Large, many grains

— Narrow, erect, thick

— Short, thick, uniculm

— Shallow

Fig. 8.1. Architectural features of a wheat ideotype.

Donald (1968). The main characteristics of this new plant type were:

- determinate, synchronous growth habit
- short, strong stem
- erect leaf habit
- few, small leaves
- large erect ear
- high harvest index
- presence of awns

Although many of these attributes are morphological, each is the result of physiological processes operating in concert during specific phenological stages (Chapters 4 and 5). Essentially, Donald conceptualized the wheat ideotype as being a poor competitor relative to its mass. In other words, each plant would be able to survive and yield at high population densities whilst making minimal demands on its neighbours for resources per unit of dry matter produced. The reproductive component, or ear, would have the capacity to absorb all the photosynthates available from other parts of the plant. In terms of phenology, minimal stem growth and early flowering would be followed by a prolonged grain filling period during which sustained photosynthesis would lead to the production of sufficient assimilates to maximize the size and weight of grains.

The salient architectural and agronomic attributes of the wheat ideotype initially devised by Donald have subsequently been modified through both theory and practice. In each incarnation, the modified ideotype remains principally designed for the production of cereal crops in benign environments with plentiful resources. At this stage, it is worth reviewing the physiological and agronomic consequences of the main ideotypic characteristics.

DETERMINATE GROWTH HABIT. The modern wheat ideotype, especially when grown as a sole crop of only one variety has greatly reduced variations in the phenological sequence of different parts of the plant caused by asynchronous or indeterminate growth habits. The introduction of a determinate growth habit and uniculm morphology means that seeds reach physiological maturity at about the same time. In this case, the production of uniform heads of comparable maturity, size, weight and composition and at a similar location above ground provides clear advantages in terms of the timing of operations, ease of harvest and the quality of the final product. Where the plant number can be assured in terms of the successful germination and establishment of each seed, the ability to tiller profusely is an undesirable attribute. A sustained sequence of tillers leads to variation in the maturity and size of grains whereas an equal population of uniculm plants produces uniformly mature grain located at the same height above the ground. A further complication of tillering is that sterile ears that contribute nothing to grain yield still use resources such as light and, particularly water, in their vegetative growth. Although some of the assimilates stored in the unproductive tillers can be translocated to the fertile ears, the net effect is a loss of resources. An additional consequence of earlier, synchronized flowering is that each plant needs to produce a smaller minimum vegetative infrastructure (w_0) before the partitioning to reproductive weight can begin (see Chapter 6).

A SHORT, STRONG STEM. The combination of shortness and strength in the stem is designed to reduce the risk of lodging with increasing weight of the more productive ears. Less photosynthate is invested in stem production and more is available for allocation to the ear, given associated increases in spikelet fertility and number of grains per ear. However, in crops such as wheat, this must be balanced against the photosynthetic contribution made by stem surfaces. The idea that the stem should be short has led to the production of *semi-dwarf* cereal varieties that have gradually supplanted their taller predecessors. The semi-dwarf morphology of modern cereals is often complemented by their shorter duration, mainly in the vegetative phase.

LEAF TRAITS. An important aspect of crop production based on ideotypes is the requirement for dense crop canopies. In such communities, nearly vertical leaves allow better illumination of a greater proportion of leaf surfaces than that within a canopy of long, horizontal or drooping leaves in which the uppermost leaves may be saturated beyond their photosynthetic limit whilst lower leaves are starved of radiation. (The theoretical and practical implications of leaf orientation are discussed in Chapter 2). As with erect leaves, a smaller number of narrow and short leaves per plant allow for a better dispersion of intercepting foliage and consequently a better distribution of radiation within dense crop canopies. Often small, short leaves tend to be more erect than longer leaves. An additional aspect is that high grain yield potential in uniculm varieties requires a high density of culms rather than a large number of leaves per culm. Under these circumstances the greater number of culms provides a greater sink capacity per unit leaf area index. The number of leaves is also associated with the degree of tillering as the axils of the lower leaves are the sites of origin of the primary tillers.

A LARGE, ERECT EAR. The maximum potential grain yield depends on the combination of the number of ears per unit area of ground and the number of grains per ear. However, Donald (1968) observed that the wheat ear was normally a limiting sink for photosynthates. In other words, the number and potential size of wheat grains, rather than the supply of assimilates was the main determinant of yield. Ear number and ear weight are usually negatively correlated, and the objective of achieving the largest possible ears must therefore lead to a decline in the number of ears per unit area of ground. The relevant objective is therefore the attainment of enough florets per unit of total dry matter to ensure that there is no sink limitation for photosynthates. To ensure the most even distribution of radiation to all sides of the ear and the canopy of leaves, an erect ear is theoretically the ideal orientation.

HIGH HARVEST INDEX. The effects of reducing stem length, tillering and the time to flowering together resulted in a marked rise in the harvest index (H). This was initially through empirical breeding trials but since the 1960s has been achieved through the introduction of dwarfing genes that reduce the ability of the stem to respond to endogenous levels of natural growth hormones such as gibberellins. The combined effects of a smaller vegetative weight at flowering (w_0) and an increase in H are diagrammatically illustrated in Fig. 8.2. Donald (1968) quotes evidence from Harland and Anthony (1920) that awns may contribute significantly to photosynthesis of the cereal ear. He

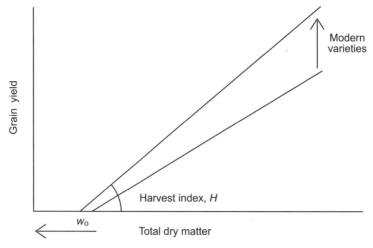

Fig. 8.2. A diagrammatic representation of the effects of smaller vegetative weight at flowering (w_0) and increased harvest index on the relationship between grain yield and plant biomass.

went on to suggest that this contribution might be relatively greater under semi-arid conditions because of the xeromorphic structure of awns compared to cereal leaves. In practice, this characteristic has not been universally adopted.

The rice ideotype

Wheat in the tropics is mainly restricted to high altitudes or the cool season and for much of the humid tropics rice is the dominant cereal. The Green Revolution in Asia is therefore most closely associated with the production of rice and, in particular, the release in 1966 of IR8, the first semi-dwarf variety of rice and its descendants. Compared with traditional cultivars, IR8 was characterized by a high-tillering, stiff-strawed, erect-leaved, photoperiod-insensitive plant that was responsive to nitrogenous fertilizer and produced a high harvest index (Chandler, 1969). The yield potential of varieties such as IR8 is typically about 10 t ha^{-1}. This, combined with the increased cropping frequency (typically three crops per year) of modern varieties, means that more than 60% of the world's rice area is planted with semi-dwarf varieties of rice (Khush and Peng, 1996).

In the 30 years since the introduction of IR8, there were marginal improvements in the yield potential of rice, with effort mainly directed to incorporating disease and insect resistance, shortening the growing period and improving grain quality. However, there has recently been a renewed effort to increase yield potential. Interestingly, much of the new thinking for rice production for the humid tropics is based on the experience of wheat in temperate regions. Khush and Peng (1996) summarize much of the recent activity in relation to increasing the yield potential of rice and have identified the following common objectives for a new rice ideotype:

- low tillering capacity (3–4 tillers per plant)
- no unproductive tillers
- 200–250 grains per panicle
- 90–100 cm height
- sturdy stems

- green, thick and erect leaves
- shallow root system
- 130 days growth duration
- disease and insect resistance
- high harvest index
- high grain quality

Many of the above characters are simply further developments of the original ideotype. However, the reduction in tillering to an optimum of 3–4 per plant is associated with the introduction of direct seeding methods due to the increase in labour costs and degree of industrial development in many rice growing regions.

Dingkuhn *et al.* (1991) calculated that grain yield of present high-yielding varieties might be increased by a further 25% through:

1. Enhanced leaf growth and reduced tillering during early vegetative growth;
2. Reduced leaf growth along with sustained high leaf nitrogen concentration during late vegetative and reproductive growth;
3. A steeper slope in the N concentration gradient in the leaf canopy with greater N present at the top of the crop stand;
4. Expanded storage capacity of stems;
5. Improved reproductive sink capacity with an extended grain filling period achieved by either or both an advance in flowering or a delay in maturity.

This modified ideotype was adopted as the 'new plant type' (NPT) identified in IRRI's strategic plan (IRRI, 1989) and has been the cornerstone of IRRI's breeding objectives for the past decade (Khush and Peng, 1996).

The limits of production and harvest index

Has breeding taken yield to its upper limit? Perhaps the single most important attribute of the modern cereal variety is its ability to allocate a large proportion of total biomass to seeds. This trend has been associated with a progressive reduction in plant height and straw weight rather than any significant increase in total biomass *per se*. Austin (1989) reviewed the historical trends in grain yields in relation to the genetic control of the maximum photosynthetic rate at saturating irradiance (P_{max}) and total photosynthetic productivity. In terms of photosynthetic rate per unit leaf area, he concluded that there was no evidence that any substantial or consistent increases in yield were the result of increased levels of P_{max}. Interestingly, there are wild diploid species of wheat that have a higher P_{max} than cultivated types (Evans and Dunstone, 1970). Although there has been no substantial increase in total biomass, Austin (1989) observed that, because nitrogen delays leaf senescence and thereby prolongs the active life of the intercepting canopy, there is an increase in the mean photosynthetic rate per unit leaf area in modern cultivars.

Irrespective of environment, the harvest index of wheat appears to be reaching its biological limit. Austin *et al.* (1989) investigated the growth and yield of 13 winter wheat varieties with dates of introduction that ranged from 1830 to 1986 with selections representing very old (1830–1907), old (1908–1916), intermediate (1953–1972) and modern (1981–1986) varieties. All varieties were grown for 3 years under conditions of uniformly high fertility and with support netting to prevent the stands from lodging. When compared with the very old varieties, modern varieties gave an average of 59% more grain, 14% more ears per unit field area, 30% more grains per ear but a similar mass per grain. Although the modern varieties produced slightly more biomass than the older varieties, it is likely that this was the result of a smaller allocation of biomass

to roots (which were not measured) rather than an increase in total biomass. The modern varieties also reached anthesis about 6 days earlier than the older ones. Over the range of varieties, there was an overall improvement of H from 0.31 to 0.53.

Figure 8.3 presents the chronological trend of above-ground biomass and grain yield for the 13 varieties tested by Austin *et al.* (1989). The slope of increasing yield clearly indicates that this must have been at the expense of other above-ground plant components rather than a substantial change in total above-ground weight. Austin (1983) also produced a theoretical calculation of what the maximum harvest index might be for semi-dwarf wheat varieties from data for four varieties. Figure 8.4a presents the actual data for the mean of the four varieties, based on a grain yield of 7.07 t ha^{-1}, stem plus sheath weight of 4.53 t ha^{-1}, chaff weight of 1.39 t ha^{-1} and leaf lamina weight of 1.43 t ha^{-1}. These data provide an average harvest index of 0.49. In Fig. 8.4b, the theoretical grain yield is calculated based on an assumption of no change in total biomass. He assumed that to support the weight of the plant, a maximum of 50% of the stem and sheath weight could be transferred to the ears with enough going to grain and chaff to retain the existing grain : chaff ratio. This theoretical minimum vegetative infrastructure produced a theoretical maximum harvest index of 0.62.

From Figs 8.3 and 8.4, it can be deduced

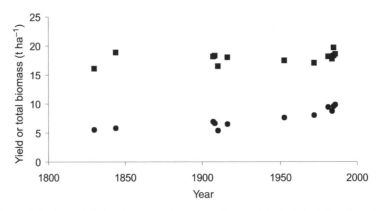

Fig. 8.3. Chronological trend of above-ground biomass and grain yield for 13 winter wheat varieties. (Adapted from Austin *et al.*, 1989.)

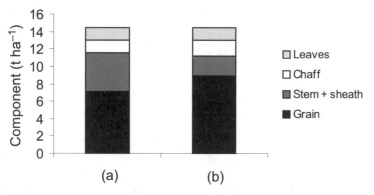

Fig. 8.4. (a) Actual data for four winter wheat varieties showing distribution of stem, chaff, leaf and grain weight. The resultant harvest index was 0.49. (Adapted from original data from Austin, 1983). (b) Theoretical calculation of maximum grain yield based on a minimum allocation to stems, chaff and leaf components. The theoretical maximum harvest index is 0.62. (Adapted from original data from Austin, 1983.)

that a harvest index of around 0.30 in about 1830 can be increased to a theoretical maximum of about 0.62. In practice, where total biomass cannot be substantially altered, the only option for an increase in grain weight is through a reduction in the weight of roots, stems or leaves. Austin *et al.* (1980, 1989) for wheat, Riggs *et al.* (1981) for barley and Fischer and Palmer (1983) for maize, all demonstrate that any varietal increase in *H* is associated with a proportional decrease in stem height. Recent evidence suggests that we are now very close to the theoretical maximum limit of *H* in temperate agriculture. For example, Spink *et al.* (2000) investigated the growth and yield of winter wheat varieties at three sites in the UK between 1996 and 1998. From their results, an *H* of 0.61 was achieved for the variety Consort. This value should not be taken as definitive evidence, nor the theoretical value of 0.62 as an absolute maximum for *H*. However, it does indicate that any further increases in *H* will need to be set against a diminishing return in terms of breeding and research effort. Of course, there is still very much more room for improvement in the actual yields of tropical cereals and the many non-gramineous crops in temperate and tropical agriculture that have as yet not achieved similar shifts in partitioning to reproductive yield.

8.1.3 The ideotype concept in harsh climates

The ideotype described above is not the sole aim of plant breeding in rice. Other rice ideotypes, designed for lowland rainfed and perennial upland rice, have longer stems, more panicles and less yield per panicle than the standard high-yielding ideotype. Nevertheless, the major crops of the world are increasingly based on a common ideotype that achieves maximum grain yield in benign or high-input environments. It is assumed that, where physical resources or biological competitors may limit crop productivity, these constraints can be overcome through methods such as the use of fertilizers, pest and disease management, and herbicides. However, there are many regions of the world, mostly in the semi-arid and arid tropics, where constraints to crop production cannot easily be removed or where inputs are not available. The conceptual basis of the common ideotype in such climates may well be inappropriate and is now reviewed.

SOWING DENSITY AND SOWING DEPTH. Because uniculm plants produce only one ear each, it is essential that planting density must be sufficient to achieve the maximum capture of resource and enough ears per unit area of ground to provide an unlimiting sink. The established population of plants must be close to the population of seeds sown. Planting a greater number of seeds than is required for the optimal established population is an agronomic option, but leads to a variable number of seedlings that may establish and use resources but never contribute to final grain yield. There is also the cost of sowing extra seeds.

Such a strategy, that assumes a uniform rate and very high percentage of seed germination and seedling establishment, might not be ideal in harsh environments because it can seldom be realized. First, there is the history and condition of the seeds to be sown. Even when seeds have been purchased from a reputable supplier or retained from the previous harvest, germinability is likely to be much lower than 100% as a result of adverse conditions in which the maternal seed was grown, poor storage environments and damage during sowing. Second, conditions in the seedbed before, during and soon after sowing play a critical role in the likelihood that seeds will germinate. Without irrigation, diurnal variation in insolation and evaporation influence the temperature and water content in the seedbed, and consequently its physical structure and degree of soil–seed contact. The temperature perceived by seeds might fluctuate to an extent that it exceeds the optimum or even the maximum temperature for germination for part of the day; consequently, the rate and fraction of seeds that germinate are reduced (see Chapters 2 and 3). In many cases, the actual time of day at which seeds are sown can influence the pattern of germination and establishment across the field. It is our own experience that seeds of millet sown in the late afternoon in a field in Niger successfully emerged and established during the night while soil temperature levels were relatively low and moisture in the seedbed was available. In contrast, many seeds sown early the next morning failed subsequently to emerge through the hot, dry soil.

There are likely to be further constraints on seedling establishment. A consequence of a semi-dwarfing habit is that the reduction in stem length is associated with a reduction in hypocotyl extension. In these circumstances, again without irrigation, the establishment of a root system in drying soil is more susceptible to physical constraints such as deep sowing and high soil bulk density and to biological factors such as pathogens. A less-than-optimum stand density is likely to ensue. In semi-dwarf, uniculm varieties, the loss of developmental plasticity provided by tillers means that the plants which successfully establish might be incapable of compensating for the space left by their missing neighbours. The problem is not simply one of reduction in the number of viable plants but also of increased spatial variability in the crop stand. For example, a reduction of, say, 30% in crop establishment is rarely consistent across the field and it is more likely that there will be pockets of almost complete germination and establishment and others where few seedlings establish.

The balance between stem and panicle

Earlier, it was argued that the continued search for an even greater harvest index in the common cereal ideotype is increasingly costly in time and effort. For tropical crops in dry environments, this approach might actually be counterproductive.

STRAW PROVIDES A BUFFER AGAINST STRESS. The reliance of modern varieties on inputs such as fertilizers and irrigation means that the carbon required for grain growth can be provided largely or entirely by current assimilation. In these circumstances, the objective of crop management is to maintain the longevity of the vegetative infrastructure of the leaves to as close as possible the final maturity of the grain. Here, the stem must remain intact and able to support the weight of the leaves and the growing ears. However, for much of the life of tropical crops, the shortage of nutrients or water in the soil means that the leaf canopy is unable to maintain its optimal function throughout the growing period. In fact, in crop varieties that have a determinate growth habit, the ability of leaves to function effectively beyond anthesis may be impaired or curtailed altogether. Where the production of current assimilates is affected by stress after flowering, grain filling must rely partly or wholly on the translocation of assimilates that were produced during vegetative growth and stored in the stems. For traditional varieties and landraces of the major crops, the contribution of stem reserves is a major determinant of final grain yield. There is considerable evidence that under stressful conditions stem reserves contribute proportionately more to the final grain weight of both modern and traditional cultivars than where conditions are non-limiting (e.g. Gallagher *et al.*, 1975; Hall *et al.*, 1990). This, coupled with the observation by Austin *et al.* (1980) that the margin between demand by the grains and the supply of assimilates from the leaves was smaller in higher yielding varieties, confirms that the reduction in stem reserves is an inherently risky option in variable environments. In fact, the relative gains in the yield potential of modern varieties have been greatest in the most favourable environments. For example, Cox *et al.* (1988) found that between 1919 and 1987 there was an overall increase in wheat yields of 41% in favourable environments compared with 27% in water stressed environments.

STRAW IS AN ASSET NOT A NUISANCE. Implicit in the pursuit of high harvest indices is the assumption that straw is of less value than grain or even of no value at all. Whilst there are circumstances in the industrialized economies where straw is an inconvenience that simply has to be disposed of without any practical benefits to the grower or the system, this is not the case in tropical cropping systems. There are numerous examples where straw from cereals and residues from other plant types can be used either in some practical capacity or for economic returns. In particular, straw from the taller traditional varieties may provide material for animal fodder, fencing, housing, soil surface mulching and incorporation within construction materials. Where these contribute to the economics of the system, the H of a crop is effectively unity. Where they have some *proven* practical value in a subsistence

system, the farmer's preference for the multipurpose value of tall varieties may outweigh the *possibility* of any yield advantages in the semi-dwarf alternatives. An associated consequence of the semi-dwarf high-yielding variety is that the stems tend to be high in lignin which along with cellulose contributes to the lodging resistance of the plant. This means that there is a reduction in both the quantity and quality of livestock feed when traditional cultivars are displaced by modern alternatives (Gill, 1991).

COMPETITION FROM WEEDS. Two central features of the cereal ideotype, lack of tillering and an erect canopy habit, may be serious disadvantages in relation to competition with weeds. Where the competitive effects of weeds can be suppressed by the use of herbicides, the vulnerability of the crop is reduced. However, where the application of herbicides is not possible, the risks of yield loss or complete crop failure are greater for the cereal ideotype than its predecessors. Interestingly, one of the options to counter the increase in the evolution of herbicide resistance in temperate cereals (Moss and Rubin, 1993) is to use more leafy, competitive cultivars (Morgan, 1989).

FLEXIBLE RESPONSE TO WATER DEFICITS. There is no greater constraint that exemplifies the lot of the subsistence tropical farmer than drought. The optimum plant population depends on soil moisture availability (see Chapter 6) and when rainfall or soil moisture varies from season to season, the ability of plants to tiller provides a means of self-adjusting the population of culms to the supply of soil moisture. Where the crop is composed of uniculm individuals, the planting density is critical to the final grain yield and as rainfall is rarely the same each year, the plant population required for each season cannot accurately be predicted at sowing. The ability to tiller progressively through the season, achieving some sort of balance between supply and demand for water, is therefore one of the most valuable adaptive response to hostile environments. Azam-Ali *et al.* (1984) demonstrated that the ability to tiller in a variety of millet growing under rainfed conditions in West Africa meant that the crop grown at 2.9 plants m^{-2} outyielded the same variety sown at 11.5 plant m^{-2}. Tillers contributed 75% of the total weight at the lowest planting density, only 21% at the highest. Although translocation of stem reserves into grain was not specifically measured, the increase in grain weight from anthesis at high density was associated with virtually no change in the overall weight of shoots. At low density, panicle and shoot weight continued to increase until close to the final harvest.

Where water is the limiting resource, an ability to root deeply and extract a larger share of the most limiting resource is clearly a further adaptive advantage in drought-prone environments (Chapter 3). Again, the *possibility* of an increase in grain yield under benign conditions must be assessed against the *likelihood* of yield losses in adverse conditions. Of course, a genetic predisposition to a large root system restricts the options both for above-ground components of vegetative and reproductive growth. Therefore, where there is no increase in total biomass, large roots must decrease both the capacity of the plant to achieve high grain yields and the option for it to transfer assimilates from stem reserves to grain filling during post-anthesis stress. For their contribution to be beneficial in drought conditions, deep rooting varieties must facilitate the production of higher yields under stress conditions through a greater total biomass than shallow rooting alternatives.

CROP DURATION AND DETERMINACY. In the common cereal ideotype, the maturity of seeds and the death of the plant are synchronized. Resources are not diverted from seed filling to vegetative growth and the yield of the single crop harvest is maximized. This ideotype, designed to obtain several crops per year in high-input systems (e.g. rice), also commonly fits the purpose of achieving one significant yield per year in arid, rainfed systems where there is a unimodal rainfall pattern of variable duration. However, there are circumstances in the semi-arid and arid tropics, where a reduction in crop duration

to avoid drought stress is not always desirable. For example, tropical climates that exhibit a bimodal rainfall pattern often provide two extended periods of soil moisture interspersed by a period of drought (see Chapter 5). The 'long' and 'short' rains of parts of East Africa provide an early growing period in which a full crop may be grown followed by a second short season which may not be adequate to support the growth of a crop from seed. An appropriate option here is to grow a cereal crop, such as sorghum or millet, in the first season and allow the ratoon crop to grow and produce a second yield within the same year and from the same root system. In these circumstances, a smaller but none the less useful yield of grain can be achieved from the ratoon crop. Alternatively, the growth of the vegetative shoots of the ratoon can be used to provide fodder for ruminant livestock.

A further, inevitable consequence of the cereal ideotype is that it is ill-designed to fit into complex canopies, such as intercrop systems. One of the major biological and management objectives of intercropping is to displace the demand for environmental resources and labour so that the peak demands for each species do not coincide. A risk of reducing the duration of one crop species is that its resource demands conflict with that of a companion species. Reductions in plant height and crop duration reduce the temporal and spatial complementarity of intercrops (see Chapter 6) because plants from different species increasingly occupy similar niches for resource capture. An example of this trend is the reduction in height and duration of pigeon pea in modern cultivars. The introduction of short duration pigeon pea designed for maximum pod yield means that these varieties mature in the same growing period as cereals at the same location. By contrast, older varieties that take up to 18 months to mature are often grown within intercrops. The relatively low rate of seed production in the pigeon pea crop is complemented by the production of stems that can be used as fuel, fodder or building material and by the production of the associated annual crops during the first season.

Selecting in the appropriate environment

Figure 8.5 demonstrates the generalized, theoretical relationship between yield and stress for modern varieties and older landraces. The objective of management is to select the most appropriate crop type to grow above or below the crossover point on the environment axis. When analysing data to establish the crossover point it is important that the range of stress is sufficiently great for both crop types. It is often the case that the highest yielding varieties may continue to maintain their relative advantage over older varieties or landraces even under moderate stress and therefore the slope of the relation is difficult to establish until a sufficiently wide range of environments has been generated (Blum, 1996). From Chapter 7 it will be clear that to generate the type of response plane required for Fig. 8.5, soil moisture availability and atmospheric moisture deficits need to be quantified, yet many varietal trials make no attempt to quantify either the soil moisture or the atmospheric deficit.

Often, the reliance on a single environment (usually non-limiting) to test varietal responses results in misleading conclusions. Ceccarelli and Grando (1991) argue that the selection of the most appropriate varieties for stressful environments should be done by comparing varieties in those environments rather than under non-limiting conditions. They compared barley geno-

types in 'low yielding' and 'high yielding' environments, defined by whether the grain yield of all the breeding lines in that environment was one or more standard deviation lower or higher than the average grain yield across all environments. They found that genotypes selected at low-yielding sites in a given cropping season had higher grain yields in low-yielding sites in the subsequent cropping season than genotypes selected from the high-yielding sites. Despite the detailed botanical information in this literature, there is virtually no analysis of environmental or physiological factors other than total rainfall; therefore, it is difficult to explain the causes of yield variation among genotypes. Interestingly, most genotypes that performed best in the low yield environments were closely related to landraces. Ceccarelli and Grando (1989) observed that the selection of locally adapted germplasm under low-yielding conditions was 28-fold more efficient than the selection of germplasm unrelated to local landraces under high-yielding conditions.

There is more widespread evidence from conventional breeding (Hurd, 1971), physiological studies (Hsiao, 1982) and quantitative genetics (Falconer, 1990) that breeding programmes to increase yield under stress should rely on selection in stressful environments. Accumulating evidence also confirms that the ability to provide a stable, though small, yield under stress and a high yield under optimal conditions are often mutually exclusive (see Rosielle and Hamblin, 1981; Ceccarelli and Grando, 1991). Breeding strategies that change the allocation of resources within the plant may not be the best solution to buffer the components of yield in drought-prone environments (Hurd, 1971; Fischer and Wood, 1979). Nageswara Rao *et al.* (1989b) demonstrated that characteristics which conferred drought resistance in groundnut required a sacrifice in yield potential under more favourable conditions.

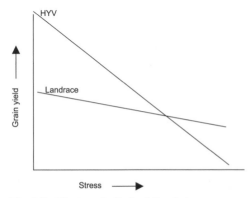

Fig. 8.5. The hypothetical relation between crop yield and environmental stress for modern high yielding varieties (HYV) and older landraces.

8.1.4 Limitations and wider impacts

The Green Revolution technology and the

development of high-yielding cereal ideotypes have been of benefit to many millions
of people over the last 50 years. The very fact
that global food production has largely kept
pace with the demands of an increasing population is largely a credit to its advocates
and practitioners. However, it has not been
successful everywhere and its adoption has
resulted in a loss of diversity, both in crop
species and forms of management. There are
significant lessons here for future development.

Constraints in environment and infrastructure

More than 80% of humanity lives in developing countries, many of which are in the
arid and semi-arid tropics. Many people living in these regions have seen little material improvement in their food output or living standards over the last few decades.
More specifically, much of Africa has gained
nothing from the Green Revolution.

Although the causes of the discrepancy between food production in Africa and
the rest of the developing world are complex, a number of phytogeographic, climatic and economic factors have been impli-
cated. Other regions in the developing
tropics each contribute at least one major
species to global agriculture; for example,
wheat from Asia, maize from South America
and rice from Indo-Burma. In most
instances, these genetic resources have been
developed and disseminated by the location
of CGIAR Institutes in or near the centres of
diversity (Chapter 4). Despite its wide range
of native crops and its wide distribution of
climates and soils, Africa has contributed
little to global agriculture. It has therefore
been largely reliant on the selection and
adaptation of germplasm from elsewhere in
the world for the bulk of its agricultural production in recent decades.

Climatic constraints to the growth of
plants (Chapter 1) are severe in much of the
continent, but equally important has been

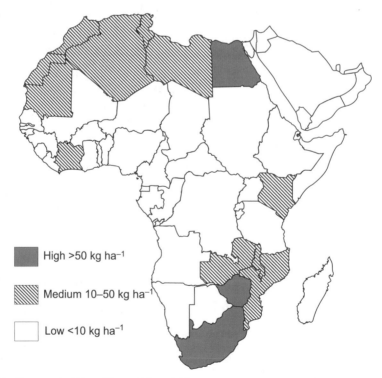

Fig. 8.6. Fertilizer use in Africa. (From FAO, 1984.)

High >50 kg ha^{-1}

Medium 10–50 kg ha^{-1}

Low <10 kg ha^{-1}

that crop improvement programmes at many of the international centres have, until recently, focused on selecting crop varieties of particular species that are highly responsive to fertilizers. This, in turn, requires either a local means of producing fertilizers on an industrial scale or their importation to the country of production. Figure 8.6 shows the distribution of fertilizer use in Africa in 1983 – at the peak of the Green Revolution in Asia – indicating those countries in which the average use of fertilizers is rated as high (above 50 kg ha^{-1}), medium (10–50 kg ha^{-1}) or low (less than 10 kg ha^{-1}). With three exceptions (South Africa, Zimbabwe and Egypt), all other African countries used, on average, less than 50 kg ha^{-1} of fertilizers and most used less than 10 kg ha^{-1}. The basic requirements to benefit from Green Revolution technology were therefore lacking in much of Africa. The net effect of these constraints is that *per capita* food production in Africa continues to decline relative to that in Asia. Figure 8.7 shows this trend for the period between 1974 and 1976 and 1985. The result has been an increasing gulf between the 'haves' and the 'have-nots', both within Africa and between Africa and the rest of the developing world.

There are arguments that the subsistence sector in Africa, and similar farming systems elsewhere in the world, should attempt to adopt alternatives to Green Revolution technology, either through their own efforts or with the assistance of sponsors. Are there alternative strategies and species which, with improvement and evaluation, could increase crop production in these hostile environments, where the use of fertilizers and irrigation is never likely to be a widespread option? One possibility is to identify the potential of those underutilized indigenous species and management systems that are best suited to the natural constraints imposed by local climates and soils. This prompts a consideration of the trends in diversity of crops and systems.

The diversity of agroecosystems

The diversity of agroecosystems can be considered in terms of the diversity of crop

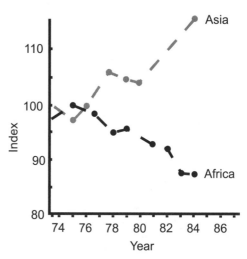

Fig. 8.7. Per capita food production in Africa and Asia between 1974 and 1985. The index is set to 100 in 1974 and relative changes are shown against this value. (Various FAO production yearbooks.)

species and the range of cropping systems. In Chapter 6, examples were given of the range of cropping systems in tropical agriculture. These span the highly diverse (agroforestry) through to the highly uniform (sole cropping or monocropping). The trend within modern agroecosystems has been towards a greater reliance on crop species and cropping systems that are more uniform, i.e. less diverse or complex than previous agroecosystems.

On a global scale, there are between 300,000 and 500,000 higher plant species of which approximately 250,000 have been identified (Wilson, 1995; Collins and Hawtin, 1999). Of these, about 30,000 are edible and approximately 7000 have been cultivated or collected by humans for food. However, as noted in Chapter 4, fewer than 20 major species account for more than 90% of global food production. Just three crops – rice, wheat and maize – account for more than 60% of the plant-derived energy in the human diet (Collins and Hawtin, 1999). The major world crops can be divided into four groups based on their product: cereals and root crops, grown predominantly for their carbohydrate content, legumes that provide protein and, in some cases oil, and those crops grown largely for the oil content in their seeds or fruits. As just described, much of the breeding and agronomic effort has been concentrated on a few major crop ideotypes within a few of these dominant species.

This intensive evolution of crop types has undoubtedly had effects on the range and diversity of crop species and cropping systems. The systematic collecting of some plant genetic resources has been very effective. There are more genotypes and landraces available now for analysis and potential use than at any other time in human history. However, the numbers of genotypes actually in use, and the methods of growing them, have without doubt narrowed. This decline in the diversity of used crops and cropping systems has displaced indigenous species by more favoured exotics. Researchers and their sponsors have sometimes led and sometimes followed this trend but have rarely provided alternatives to it, by assessing the potential benefits, whether in terms of yield or sustainability, of alternative or indigenous species.

Two sets of questions are generated from the convergence of much of agriculture towards a single method. One is the effect on wider biological diversity. The other is the reason why so few genotypes provide such a great proportion of our food. Are there good reasons why other crops are not grown on a world scale or is it simply because the full potential of many hundreds of underused species have not been scientifically assessed and analysed?

Within any particular agricultural species, the extent of its genetic diversity is critical to its future agronomic potential. It is this variation that enables the species to adapt to new agroecosystems and environments through natural and human selection (Collins and Hawtin, 1999). We can consider the range of species grown and the genetic diversity within each species in the context of agricultural research and extension. Most of the major staple crops of the world are 'mandate' species of the various CGIAR centres (see Chapter 4). This provides a global platform and an international respon-

sibility for research, extension and germplasm evaluation for these species. Even within this group of 'favoured' crops we can see the disparity between those crops grown in the richest and those grown in the poorest regions of the world, particularly with respect to the protection and expansion of their genetic resources. For example, Cassava provides over 50% of the plant-derived dietary energy for Central Africa but only 1.6% of that globally (FAO, 1996). By contrast, wheat provides the basis of the diet in the most developed temperate regions of the world but is virtually non-existent in much of Africa. It is interesting to note that whilst globally there are an estimated 784,500 germplasm accessions of wheat, the comparable figure for cassava is 28,000 (FAO, 1996).

For crops not protected by a CGIAR mandate, the situation is even bleaker. Again, FAO (1996) notes that unless there are well-developed crop-specific networks, the lack of attention paid to these underutilized species leads to an absence of information and lack of any oversight and monitoring responsibilities. It is noticeable that many of the species not adopted by the CGIAR have their centres of diversity and production in the semi-arid and arid tropics, most noticeably in Africa.

The absence of any overarching strategy or infrastructure for these alternative or underutilized crops has implications for the protection and enhancement of their genetic diversity. Later in this chapter we consider the example of one such species – bambara groundnut – for which there is no CGIAR responsibility. Despite its wide distribution across a range of agroecosystems and its importance to the subsistence diet of many Africans, there are only 3500 germplasm accessions of bambara groundnut worldwide. Where there is no international obligation to protect and promote a species; these responsibilities must be undertaken by national agencies. It is ironic that the responsibility for many underutilized, indigenous food crops is left with national agencies within some of the world's poorest nations – often the very organizations and nations least equipped to fulfil this task. With this in mind, it could be argued that it is major crops such as wheat, rice and maize that least need the patronage of the CGIAR network.

IDENTIFYING ALTERNATIVE SPECIES. An underutilized species is, by definition, one that has not been fully exploited in terms of its potential use as a food or non-food product. There are many reasons why potentially useful crop species do not attract interest from agricultural science. Of these, the principal constraint is often lack of funds to carry out research because potential sponsors are prone to ask why 'if a crop has so much promise has no-one else funded research on it before?' Another argument from sponsors is that even a small increase in the average yield of a major species, e.g. rice, will have a massive global impact because of the area under cultivation, whereas significant increases in the yield of a minor crop can, at best, have only a local effect. On the other hand, it is impossible to know whether any underutilized species has a global potential unless it can be evaluated under a range of limiting and non-limiting conditions.

Inevitably, such research is expensive and manpower-intensive and clearly it is unrealistic to investigate every possible food crop. Therefore, before advocating the use of scarce research funds on any particular underutilized species, it is appropriate to determine:

1. Which underutilized species should be selected for investigation;
2. What research strategy should be adopted;
3. How research results and recommendations can be disseminated to the end user.

The next main section of this chapter describes how these questions were approached for one African crop species. Before that, two technological developments are briefly discussed that are likely to affect all future progress in improving crops and cropping systems.

8.1.5 New global technology

The quantitative methodologies that set the direction of agronomy introduced computerization to a receptive body of researchers, while plant improvement rapidly adopted genetic knowledge at the molecular scale to serve its aims. Agronomic research can benefit considerably from both these developments.

Information technology: 'knowledge engineering'

One of the major issues facing developing countries is whether they will benefit from the expansion of global information systems or be left further behind as a result of them. Information can now be readily transferred around the world between research groups as basic data, as descriptions of technique and as models that mathematically describe some link between environment, practice and yield. One of the main issues in practical, tropical agronomy is the value of computer-based models of tropical crops and cropping systems.

The multilocational trials required to identify the input–yield curves at each location are time-consuming, labour-intensive and expensive. Also, it is extremely difficult to extend the validity of such trials to different crop species or to new areas and to future combinations of weather variables. Modelling offers one solution: to reduce the timescale required for research results to be useful and to extend these results to new environments or species without the need for repetitive effort. Crop models organize information collected from numerous experiments into more general rules of crop behaviour; they work through mathematical statements of how a crop behaves in response to particular factors, such as light, water and fertilizers. Because these statements are mathematical, they have to be explicit at every stage and must rely on particular assumptions about the crop in question. If modelled predictions are not an accurate description of reality we are forced to accept that the system in question is not fully understood and needs further investigation.

EMPIRICAL AND SEMI-EMPIRICAL (INDEX) MODELS. There is a general divide between empirical models and mechanistic, or simulation, models. Empirical models are primarily based on relations derived from regression analysis, usually of large amounts of data. The disadvantage of this type of model is that (as for yield–input functions) it cannot be applied outside the region for which it was developed. Despite these limitations, this type of model has its advantages. For example, at the site at which it was developed, an empirical model may provide more accurate predictions than a simulation model. This type of model can therefore be a very useful tool, provided its use is restricted to its region of calibration.

Beyond the simplest empirical models are a large group of practical tools which can be classified as 'index' models. They do not simulate crop processes but empirically relate crop performance to an appropriate index, which varies with the limiting environmental factors. A typical index is the 'crop factor' used in some procedures for estimating transpiration or water requirements. The potential evaporation measured or calculated for a location is multiplied by the appropriate index to get actual evaporation from the field. The index itself is obtained from a large set of data. For instance, canopies of some species are often shown to need more water than others to grow at the maximum rate. The index might change in a reproducible way as the canopy expands; and some species might habitually lose more water than others when growing on a drying soil. The index is probably more transferable than simple empirical regressions, but still has (or should have) major restrictions on its use.

SIMULATION OR MECHANISTIC MODELS. In this type, there is an attempt to base the structure of the model on real processes that take place in the system. Many crop simulation models are based on the concept that a crop captures resources, such as light, water and nutrients and converts these into biomass or plant products via the process of photosynthesis. The captured resources are then partitioned between various organs or lost through respiration and senescence. Developments in crop simulation modelling have taken advantage of the fact that the life of all green plants is mediated by universal mechanisms. The structure can be transferred between crop species. Once a model has been worked out for one type of plant, it can readily be worked out for another.

At the hub of any model is a 'growth engine', which provides some means of calculating the amount of product the crop will produce in photosynthesis. The process of photosynthesis is extremely complex but ultimately involves the reduction of CO_2 to form molecules such as glucose using absorbed radiant energy or light. Glucose molecules produced via photosynthesis either provide the building blocks for virtually all the organic constituents of plants or are respired to provide energy for metabolic processes. An important feature of the photosynthetic pathway is that, at the same time as CO_2 is absorbed by plant leaves, they must also lose a fixed amount of water via the process of transpiration. Clearly, there are special circumstances in which either the supply of water or the interception of solar radiation alone dominate the process of dry matter production by crops. However, in most cases, crops are continuously responding to short-term changes in both solar radiation and water supply, neither factor being optimal. Consequently, in order to successfully predict the performance of crops throughout their growth and across the widest possible range of circumstances, robust crop models should take account of both light- and water-driven productivity. By comparing the predicted or modelled response of the crop against the actual response observed under experimental conditions, the model can be validated for these specific conditions.

SUMMARY. Crop yield models have a number of practical applications. They can be used to:

- reduce the need for site-specific long-term field experiments
- interpret weather records in terms of pro-

duction potential and climatic limitations

- evaluate expected returns to soil and crop management practices
- assess risks associated with management practices
- communicate research results between locations

In many cases, models are also appropriate as *relative* rather than *absolute* tools. For example, a model might be used exclusively as a means of indicating whether a given management option is better than an alternative practice – the exact magnitude of the increase may be of secondary significance but the grower at least knows which is the best option in a given situation. Nevertheless, great care is needed to appreciate the assumptions in the model and the conditions in which it can reliably be used. Too many models are a 'black box', require too many input variables and are unable to incorporate variability in physiology or environment. Examples of more flexible and user-friendly models are briefly considered in the final part of this chapter.

Molecular techniques in crop improvement

Recent developments in the basic biological sciences and their associated technologies offer opportunities for further genetic improvement of crops and for increasing our knowledge of crop science, production and management. Scientists in many areas such as agronomy, plant breeding and quality control are encountering and using such information with increasing regularity. The techniques of molecular biology are being used in two principal ways: to define a more certain link between the genome and a trait of agronomic interest; and to generate new combinations of genes and traits that would be unlikely to occur by conventional plant breeding. Forms of the latter use are considered later in the chapter.

Methods of estimating genetic diversity and identifying superior individuals in crop improvement programmes have relied upon statistical analysis of morphological and agronomic traits. Distinct traits, that are always expressed by a genotype, are used as

markers in breeding programmes. Such traits might be a particular form of branching, or a particular shape of panicle. They indicate that the genetic information coding for the marker is present in the plant in question. However, many morphological characters may not be significantly distinct, usually require growing the plants to full maturity before identification, and are strongly influenced by the environment. Some molecular techniques circumvent or reduce these complications by directly defining genetic sequences in a plant's DNA that either determine or are in some way associated with a phenotype.

The available techniques differ substantially in the degree of association between genes and trait. The DNA sequence of the small temperate cruciferous weed, *Arabidopsis thaliana*, has now been determined, and that of rice is soon likely to be. Even with this information, the links between DNA sequence and plant form and function are clearly not direct, but such information opens up the scope for defining which DNA sequences give rise to the functional proteins operating during specific phases of development and growth.

A more general class of molecular techniques, in widespread use today, involves markers or DNA sequences that probably have no direct or explicit role in gene functioning. Rather they give evidence of the genetic similarity of two or more plants, varieties or landraces, from which might be inferred something of their evolutionary origin. The limitation of many of these techniques is that the patterns among individuals are associated uncertainly, and sometimes weakly or not at all, with the phenotype (Bachmann, 1994). For instance, considerable phenotypic variation has been reported even in species with a narrow genetic base such as groundnut, while the use of molecular markers shows apparently low molecular variation among genotypes in this species (He and Prakash, 1997; Singh *et al.*, 1998). More useful in breeding programmes are DNA markers that can readily be detected, that are linked or associated with a specific phenotypic trait and whose inheritance can be monitored (Ford-Lloyd and Painting, 1996; Newbury and Ford-Lloyd, 1999). Nevertheless, molecular tools have become widely used for a range of purposes including: genetic diversity studies, for example comparing crop varieties with wild types; identification of genetic loci associated with the expression of agronomic traits; marker-assisted selection to speed up the breeding process; establishing varietal identity, purity and stability; determining pollen donors and parentage; and screening and selection of germplasm.

Molecular technology should interact with agronomy in much the same way as conventional plant breeding has interacted with it to raise yield or improve the quality of yield. Additionally, it should provide plant material with more specific and more certain genetic differences for agronomic study. At present, the agronomic significance of physiological traits is determined by comparing varieties which have observed differences in those traits. The plant material invariably differs in other traits and may also differ in ways not fully appreciated. More precise information on 'downstream' implications for water and radiation flux could be obtained by comparing plant types having one, and only one, genetic difference pertaining to the function being examined.

8.2 Assessing the potential of underutilized crops: a case study of bambara groundnut

Research on bambara groundnut is a rare example of an international effort to study and assess the potential of an underutilized crop by applying a range of scientific disciplines. Since 1988, scientists in four European countries, viz. United Kingdom, the Netherlands, Germany and Italy, have joined forces with scientists, growers, traders and consumers in Botswana, Tanzania, Sierra Leone, Swaziland, Namibia and Zimbabwe to work on this indigenous African legume. Together, their efforts provide a comprehensive assessment of the eco-physiology, agronomy, nutritional biochemistry, agroprocessing, genetics and marketing potential of bambara groundnut. Though all

of these activities were not always collaborative or systematically structured, the experience provides a basis to consider whether a similar approach can be used for rapid evaluation of other under-researched species. Further details of the material presented below appear in Azam-Ali *et al.* (2001a). Here we summarize how a methodological framework might be developed, using bambara groundnut as a case study.

8.2.1 How can an underutilized crop be evaluated?

In most developing countries, the time available for research is limited and funds are scarce. In these circumstances, it is worth asking whether precious financial and human resources should be dedicated to research on any underutilized species about which little evidence has been systematically gathered or published. It may be that many of these species have already been investigated, but they have failed to compare with the nutritional or economic qualities of the species that supplanted them. This may even be so in those regions in which the underutilized crop arose, diversified and began to be cultivated. The questions to ask about any underused food crop are:

- Has its recent history contributed to its restricted use?
- Does it have food value or economic value?
- What is the potential yield of the species?
- Can its geographical range be mapped?
- What is its genetic diversity?
- How can its existing germplasm be improved?
- Can experience gained on this crop be extended to other underutilized species?

8.2.2 Recent history and development

Bambara groundnut (*Vigna subterranea*) is an indigenous grain legume, grown mainly by subsistence women farmers in drier parts of sub-Saharan Africa. Its seeds can be eaten fresh when semi-ripe, as a pulse when dry and mature, or can be ground into flour (Linnemann and Azam-Ali, 1993). Its common English name derives from the Bambara tribe that now lives mainly in Mali. However, its centre of origin is thought to be from the Jos plateau and Yola regions of northern Nigeria through to Garua in Cameroon and possibly as far as the Central African Republic (Hepper, 1963). For many centuries bambara groundnut has been cultivated in tropical Africa, south of the Sahara. However, slaves also took the crop to Surinam and it has been found in small quantities in South and Central America, India, Indonesia, Malaysia, the Philippines, Sri Lanka and parts of northern Australia (Linnemann and Azam-Ali, 1993).

The first description of bambara groundnut in the international literature appeared early last century (Zagorodsky, 1911) but it was not until 1969 that a detailed analysis of the growth habit and reproductive behaviour of the crop was published (Doku, 1969). However, much of the literature on the crop has been specific to particular countries or has addressed only one particular aspect, such as the nutritional value of seeds (Poulter and Caygill, 1980) or visual descriptors of the germplasm (IBPGR, 1987). There is very little literature that describes the responses of the many different landraces to various cultural practices across a range of environmental conditions, nor has there been any integrated effort to improve the existing germplasm through breeding, seed selection and multiplication, or agronomic management (Linnemann and Azam-Ali, 1993). The breeding system remains unknown and no one has yet successfully produced a true variety of the crop. Local expertise and germplasm have rarely been exchanged between different growing regions in Africa, nor has knowledge been recorded and widely disseminated, either informally or through the scientific literature. Nevertheless, over many centuries, farmers have accumulated a wealth of knowledge on the crop and cultivated an array of landraces that have become well adapted to the local climates and soils. In much of Africa, bambara groundnut is the

third most important legume after groundnut and cowpea (Sellschop, 1962).

The causes of the crop's neglect and decline compared to these other legumes are unclear. Limited evidence indicates its tolerance to adverse environmental conditions, yielding on soils of low fertility and with little rainfall. The seeds often command a high market price, demand far outweighing supply in many areas (Coudert, 1984). However, there is a widespread perception that bambara groundnut has limited economic potential outside its areas of cultivation. For example, groundnut (*Arachis hypogaea* L.), a crop that was first introduced to West Africa from Brazil by the Portuguese (Smartt and Simmonds, 1995), may well have replaced the local bambara groundnut because its seeds contain large amounts of lipid and therefore serve as an oilseed. In common with many other African species, the status of bambara groundnut as a 'poor man's crop', generally grown for subsistence rather than cash, may have played some part in its marginalization by decision makers and sponsors. Until recently, bambara groundnut, like many indigenous African crops, has been virtually ignored by international agricultural scientists and funding agencies.

8.2.3 Food value and economic demand

Recent studies on its composition (Brough and Azam-Ali, 1992; Brough *et al.*, 1993) demonstrated that bambara groundnut is a rich source of protein (16–25%) and confirmed previous evidence (Poulter and Caygill, 1980) that it has potential for food processing and an increased range of uses. The protein content was found to be superior to that of cowpea, groundnut and pigeon pea. Although the protein in bambara groundnut seeds is deficient in methionine and cysteine (a common feature in legumes), it achieves the FAO requirements for all other essential and non-essential amino acids. Brough and Azam-Ali (1992) concluded that, in terms of lipids, bambara groundnut seeds, having a lipid fraction of 7.9%, compare favourably with cowpea

(1.0–1.6% dry weight) and pigeon pea (1.2–1.5%) but not with groundnut (45.3–47.7%). The gross energy value of bambara groundnut seeds is greater than that of pigeon pea, cowpea and lentil. Bambara groundnut seeds can be used to produce a vegetable milk that is comparable with soy milk and the ground seed can compete with or replace other conventional flours in a range of processed products (Poulter and Caygill, 1980; Brough et al., 1993). Sensory analysis showed bambara groundnut milk was ranked first when compared with milks prepared from cowpea, pigeon pea and soybean (Brough et al., 1993).

Given this high nutritional value, the next step is to assess whether products made from its seeds can fetch a price that would meet the demands of a local or international market and justify the infrastructure required for transport and distribution. To provide an independent assessment of market potential, the Crop Post Harvest Programme (CPHP) of the UK Department for International Development (DFID) funded work, starting in 1999, to investigate the factors that affect the processing and increased use of bambara groundnut in two sub-Saharan countries: Zimbabwe, which had a relatively developed infrastructure; and Swaziland which had a smaller, less developed economy (Hampson et al., 2000).

A market survey was designed to evaluate the continuum from production through to marketing of the crop and to analyse factors limiting its increased utilization. The market survey was instrumental in identifying and recommending potential areas for further development and research. The project was split into three phases: preparation in the UK; field work and data collection in Zimbabwe and Swaziland; and data analysis, writing and dissemination of reports. Published and unpublished or 'grey' literature on bambara groundnut in Africa was collected, specifically in relation to Zimbabwe and Swaziland. BAMNET (an international network of scientists and institutions with an interest in bambara groundnut (http://www.dainet.de/genres/bambara/)) was contacted and an outline of the project

was posted on the BAMNET Web page.

In both countries, the survey methodology relied on individual and semi-structured group interviews of informants selected with assistance from local agricultural extension officers. The survey was split into categories that included as respondents farmers, traders, consumers, wholesalers, government institutions, non-governmental organizations, educational and research institutions. Checklists were used to group and code qualitative information into matrices which were analysed using the qualitative software analysis programme, QRS NUD*IST 4 (Qualitative Solutions and Research Pty Ltd, Australia). The main findings were:

- The market comprises new seed for eating fresh and dried seed for cooking and sowing.
- The market is mostly informal, there being few stages between farmer and consumer.
- Most of the production is from small-scale farms and holdings.
- Little of the crop is processed to give a value-added product.
- The bulk of the local market consists of seed cooked from fresh as a snack with few known recipes for the dried seeds, which are hard and take a long time to cook.
- The export market is presently small, demand outstripping supply in many areas of southern Africa, but the potential for export of dried bambara groundnut is high, as seed for planting or as food.
- Constraints to the increased use include a lack of awareness of the best agronomic practices, the large amount of fuel energy required for cooking or processing, lack of recipes, ignorance of the crop's nutritional value and its limited potential as a cash crop, the availability of seeds and the access by farmers to markets.
- Recommendations were to include farmers in agronomic trials, the development of recipes and processing options to reduce cooking time, development and implementation of an extension package for small-scale farmers, marketing and

promotion of final products by retailers, establishing market links and support mechanisms for promotion and exploring the possibility of new export markets such as Fair Trade or organic niche markets (Hampson et al., 2000).

Despite the constraints to its use and marketing, 98% of farmers interviewed in another recent survey regarded bambara groundnut as a profitable food crop and are keen to increase its productivity (Sesay et al., 1999). It is important to mention that the constraints and limitations encountered during the market assessment of bambara groundnut are largely similar to those historically identified before the development and expansion of most crops. Unlike bambara groundnut, however, comparable species such as cowpea, pigeon pea, groundnut and lentil have all benefited from promotion and advocacy by the network of international research centres, governments and international donors.

8.2.4 The agroecological potential of the species

First estimates of potential

Since 1988, the University of Nottingham, UK has conducted research on bambara groundnut in response to requests from a number of institutions in Africa. The methodology was derived from previous research on subsistence tropical crops, which aimed to establish relations between climatic factors and the growth of tropical cereals and legumes. An essential feature of these earlier studies was that measurements on crops growing in five purpose-built glasshouses at Nottingham (Sutton Bonington) should complement field experiments in the tropics. The glasshouses provide a means of accurately controlling environmental factors, such as air temperature and humidity, by the use of a central computer, whilst growing crops with their roots in undisturbed soil. Thus, the effect of a single variable can be monitored and quantified at the crop level so that specific factors that limit crop productivity can be identified under closely controlled

conditions (Monteith et al., 1983).

In 1986, the University of Nottingham established the Tropical Crops Research Unit (TCRU), which provides a base for a broad range of multidisciplinary activities on tropical species that grow in hostile, marginal environments and for which the processes that limit productivity are poorly understood. The glasshouses that had formed the basis of the previous activities were refurbished with improved heating, CO_2 injection and environmental control software. In 1988 and 1989, stands of bambara groundnut and groundnut were compared across a range of soil moisture levels and confirmed much circumstantial evidence that bambara groundnut is more drought tolerant than groundnut and can produce a yield of pods in environments where groundnut may fail completely. It was envisaged that, in future, the best use of the controlled-environment facilities was to examine effects of the global factors, temperature and photoperiod, which are difficult to assess in the field.

In 1990, experiments at Nottingham and also at the University of Wageningen in the Netherlands (Linnemann, 1994), showed that, although the flowering behaviour of many bambara groundnut landraces is unaffected by daylength, the filling of pods appears to require daylengths of about 12 hours. Although the mechanism underlying this process is still unknown, it has important practical implications and might explain the variability of bambara groundnut yields across Africa, where crops are generally sown in response to the onset of rains rather than at a specific daylength. At Wageningen (Linnemann, 1994), studies on eight landraces of bambara groundnut from Nigeria demonstrated that all were day-neutral for flowering. However, photoperiodic differences were found in the onset of podding. A 14-h photoperiod delayed or inhibited the onset of podding. This was caused by a check on the growth of ovaries that had been successfully fertilized. Where they had been produced under the 14-h photoperiod, ovaries developed into full-grown pods after transfer to a 12-h photoperiod.

In experimental conditions at

Nottingham where water was not limiting and daylength was maintained at 12 h, a bambara groundnut landrace from Zimbabwe achieved pod yields equivalent to over 3 t ha^{-1}. This value was greater than that achieved from a groundnut variety from India that had been grown under the same experimental conditions and was many times greater than typical yields of bambara groundnut achieved in the field. In principle, therefore, no gross constraints in the physiology of the species were apparent.

Defining the 'global' and ' local' constraints to production

The next step was to establish the agronomic potential and environmental constraints to the productivity of bambara groundnut in its current areas of cultivation. In 1991, the complementary research at Nottingham and Wageningen and links established with a number of African institutions resulted in the award of a major contract from the European Union (EU) that was designed to assess the potential of bambara groundnut for different environments in Africa. Tasks included identifying the physiological attributes associated with the ability to produce yields under dry conditions, constructing a crop simulation model for predicting the yield of different genotypes in contrasting environments, defining agroecological regions and seasons for the cultivation of bambara groundnut in the African countries, and recommending management practices to stabilize rainfed yields. The final objective was to outline a methodology to assess the potential of other underutilized species in tropical environments.

The main activities between 1992 and 1996 were to measure the physiological variables, defined in Chapters 2 and 3 as controlling radiation-limited and water-limited growth and yield. Research in controlled environments in Europe was intimately linked to field work on the crop in Tanzania (Sokoine University of Agriculture), Botswana (Botswana College of Agriculture), and Sierra Leone (Njala College). In Botswana and Tanzania, experiments were conducted in all four years between 1992 and 1995. In Sierra Leone, studies were confined to 1992 and 1993 due to civil conflict. The particular role of research in the controlled environments was to assess the effects of temperature and photoperiod; while at various times and locations and for contrasting landraces, experiments in the field in Africa investigated the effects of sowing date, soil moisture, planting density and other factors. A common protocol was adopted for the design and execution of field experiments and the collection and collation of crop measurements.

Some of the functional relations in the experimental data were of a general and universal nature, while others were very specific to bambara groundnut (Babiker, 1989; Zulu, 1989; Berchie, 1996; Collinson et al., 1996, 1997, 1999, 2000; Karikari, 1996; Sesay et al., 1997; Kocabas et al., 1999; Karikari et al., 1999). The values of traits defining use of radiation and water were consistent with those in similar species. For instance, the conversion coefficient for intercepted radiation (ε_s) ranged between 0.8 and 1.1 g MJ^{-1} (total radiation) across contrasting landraces. The comparable range in the dry matter/water use ratio (ε_w) was 1.8–3 g kg^{-1}. However, this range was largely accounted for by variation between treatments in the actual leaf to air vapour pressure deficit; the resultant 'transpiration equivalents' varied between 4.2 and 4.6 g kPa kg^{-1}, were therefore conservative and similar to values measured for other legumes such as groundnut.

The traits most specific to bambara groundnut were those mediating the control by daylength of podding (rather than flowering) and the consequent influence on the balance between vegetative and reproductive growth. Controlled-environment studies identified landraces that are slightly sensitive to daylength for flowering and strongly sensitive for podding, others that are insensitive for flowering and sensitive for podding and still others where a combination of temperature, photoperiod and minimum development rate best describe reproductive development. In some cases, transferring plants from 14 to 11 h day^{-1} photoperiods at various dates accelerated the rate of podding when compared with plants maintained under continuous short photoperiods. In the landraces used in this study, critical periods for podding ranged from 10.6 h day^{-1} at 26°C to 14.1 h day^{-1} at 21°C. The complex behaviour of bambara groundnut landraces in relation to daylength may help to explain large inter- and intra-annual variations in pod yields, and the uncoupling of yield from dry matter production, previously observed across locations in Africa.

Conclusions

This project provided unequivocal evidence that bambara groundnut is a crop with considerable potential deserving further investigation. Under rainfed conditions, where the sowing date was synchronized with the daylength requirement for pod filling, field crops achieved yields greater than 3 t ha^{-1}. Preliminary predictions for Tanzania indicated that pod yields in excess of 4.5 t ha^{-1} are possible in appropriate agroecological regions and with suitable agronomic management. Indeed, an ancillary experiment in Botswana on local landraces grown with supplementary irrigation identified pod yields in excess of 5 t ha^{-1} which were obtained at a planting density of 1 plant m^{-2}. From the evidence assembled during this project, a closer matching of crop production to the availability of resources during the growing season should result in higher annual yields and less variation between seasons, even without significant improvement in the existing germplasm.

Central to the attainment of the yield potential, and irrespective of all other environmental factors, is the requirement to match the photoperiodic response of podding to the temporal availability of resource during the season. This has enormous implications for the future development of the crop. An interesting corollary of the investigations so far is that it required research in controlled-environments to identify that pod filling had a daylength requirement. This characteristic had never been suspected in field studies, despite the wide range of environments and seasons across which different landraces have been grown for centuries. Moreover, the physiological approach used by the collaborating partners in this project was vindicated as a basis for investigating a promising underutilized species.

The exercise also revealed deficiencies that require attention in future efforts. The project would have been more efficient had there been agreement at the outset on the experimental philosophy and on protocols, better access to data on local environments and soils, and procedures to ensure the successful exchange of germplasm between collaborating institutions. Most crucially for future work, it became clear that local knowledge should be incorporated within the experimental programme, and that there should be much greater interaction and exchange of information between collaborating scientists and the farmers and consumers in each African partner country. Parameterization of the model would undoubtedly have benefited from better field-based estimates of conversion coefficients for water and light, growth of roots and shoots and canopy surface areas. Validation of the model was limited more by a shortage of adequate environmental data (for example, sufficiently localized records of daily solar radiation are not widely available across Africa) than by physiological knowledge. Nevertheless, the project generated the first international symposium on bambara groundnut (Anonymous, 1997) which brought together project scientists from each partner institution and scientists working on the crop in other countries.

8.2.5 Mapping the geographical potential

Despite the crop's sporadic appearance in other regions, Africa is both the centre of diversity and current production centre for bambara groundnut. However, as with any crop, there are locations outside its current distribution that may offer conditions at least as conducive to its growth and yield as those that are currently cultivated. We have already seen that bambara groundnut has not benefited from a systematic effort to dis-

perse it either through migration, colonialism or research. The problem remains of
how new potentially suitable areas for the
cultivation of any underutilized crop can be
identified without advocating a major multilocational programme of field experiments.
In addition, how can any information that is
gathered at various locations be collated to
provide a more generalized picture of the
potential distribution of the crop?

Prediction through physiological modelling

Such a predictive exercise requires a means
of generating the curves for light-limited and
water-limited growth and yield (Chapter 7)
at a large number of points on the earth's surface, given the actual weather and soil, or
some representation of them, at these points.
The only basis for prediction is the knowledge obtained in controlled conditions and
in the field, which links physical variables
of the environment to yield via the physiological processes of the plant. This knowl-

edge was obtained in a restricted range of
environments. However, it can be used in
prediction outside this range because experience tells us that traits such as the tem-
perature-driven curve of fractional interception on time and the conversion ratios are
maintained from one environment to the
next.

The vehicle chosen for prediction was
a process-based, simulation model named
BAMnut (Fig. 8.8), the first version of which
was constructed during the previous EU
project. BAMnut was based on the PARCH
(Predicting Arable Resource Capture in
Hostile Environments) series of models
under development at the University of
Nottingham and elsewhere. The model
works first by reading meteorological data
from files containing irrigation or rainfall,
maximum and minimum temperature,
atmospheric saturation deficit, pan evaporation and solar radiation. These variables
drive emergence, leaf production, flower-

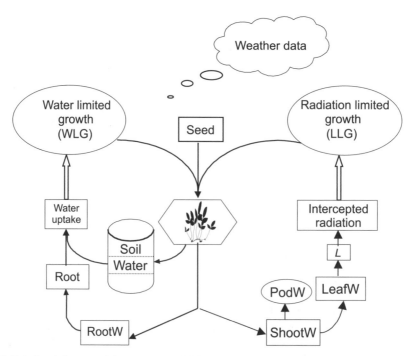

Fig. 8.8. Relational diagram of the BAMnut model. The water balance for growth under water-limited
conditions is represented in the left part of this figure. The right part of the figure represents potential
growth of shoots and pods. *L* is the leaf area index, LeafW, PodW, RootW and ShootW represent the
respective dry weights of crop components. (From Azam-Ali *et al.*, 2001b.)

ing, and other phenological traits, which themselves are back-coupled to the meteorological data to define the amounts of light and water that are 'intercepted' and 'converted' into dry matter on each day (Fig. 8.8). The actual production on any day is calculated as the lower of the light-limited or water-limited calculations. An index of crop stress is derived from the ratio of light-limited to water-limited growth. The model then partitions daily assimilation between plant organs depending on developmental stage and the value of this index. The rate of development to flowering, and then to podding, depends on daylength and temperature.

The FAO mapping project

In 1999, the Food and Agriculture Organization (FAO) of the United Nations collaborated with the TCRU group at Nottingham to map locations and areal expanses that have potential for the production of bambara groundnut across the world (Azam-Ali *et al.*, 2001b). To achieve this, project scientists linked a weather generator and a version of the BAMnut model into a Geographical Information System (GIS). The methodology established during this project was used to evaluate the productive potential in Africa, where the crop has been grown for centuries, but where experimental evidence is scant, incomplete or inaccessible; and to define new regions not previously associated with the cultivation of bambara groundnut, but where the combination of environmental factors indicate a potential for productive growth without recourse to costly inputs, such as irrigation.

For the period 1961–1990, a gridded mean monthly climate dataset at a resolution of 50 km × 50 km for global land areas (excluding Antarctica) was used as input to the weather generator to generate daily weather data. Simulation results from BAMnut were used as inputs to the GIS to provide the required maps and statistics. Given the timeframe of the study, neither the model nor the overall methodology attempted to account for the specific effects of soil type, pests or diseases on the likely productivity of bambara groundnut at any location. Similarly, any differences in daylength sensitivity for pod filling in many bambara groundnut landraces were not assessed in relation to potential yield. However, by scrutinizing the world for bambara groundnut potential over relatively small unit areas, GIS technology linked with the BAMnut model provides a means to ask:

1. Where is the best potential for growth?
2. What is the likely yield threshold?
3. Over how much surface area of land are potential yields achievable?

BAMnut needs data on the most important factors that affect crop yields – these are the model inputs. After passing 'through' BAMnut, these inputs are converted to a number of outputs, such as maps and statistics of crop yields. A schematic diagram summarizing the methodology developed in this study is shown in Fig. 8.9. The outputs of multiple simulations of the model for both biomass and pod yield have been classified into four representative ranges of suitability as shown in Table 8.1. Although these categories are arbitrary, they help to simplify the analyses and provide a basis for comparisons between regions. The levels, defined as very suitable (VS), suitable (S), moderately suitable (MS) and unsuitable (US), are based on reported pod yields for bambara groundnut at different locations in Africa. For example, the VS category with pod yields greater than 3000 kg ha^{-1} corresponds to reported pod yields of up to 3870 kg ha^{-1} in Zimbabwe (Johnson, 1968). Similarly, the MS category corresponds to the typical farmers' yields of 650–850 kg ha^{-1} in Africa reported by Stanton *et al.* (1966). Regions producing pod yields below 300 kg ha^{-1} are defined as unsuitable (category US). Crop failure has not been included as it obviously has the definition of no yield.

Analysis

Figures 8.10 and 8.11, respectively, show the predicted biomass and pod yield of bambara groundnut across the world. The different suitability ranges defined in Table 8.1 indicate the geographical distributions of

INPUTS   **BAMnut MODEL**   **OUTPUTS**

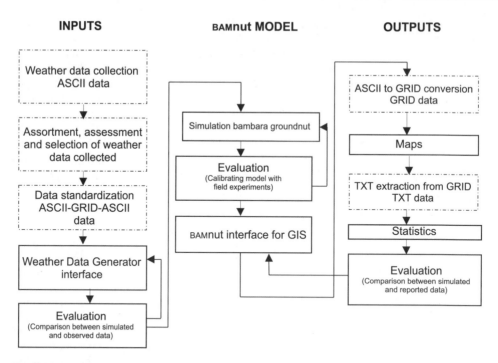

Fig. 8.9. Schematic diagram describing the procedures involved in estimating and mapping the potential production of bambara groundnut. Note: Dashed boxes refer to GIS activities while solid boxes refer to the modelling activities. (From Azam-Ali *et al.*, 2001b.)

each classification in Figs 8.10 and 8.11. There is potential for bambara groundnut production in many parts of the world beyond its current distribution, there being suitable areas with potential in America, Australia, Europe and Asia as well as Africa. In fact, locations within the Mediterranean region show the highest predicted biomass, often exceeding 8500 kg ha⁻¹.

This study used only data that are comparable across the world. The approach limited the number of important factors that could have been used in the evaluation (e.g. soil data), but it enabled comparisons to be made between countries based on consistent climate data. Predictions suggest that much

of the Mediterranean basin appears to provide the ideal agroecological conditions for bambara groundnut, potential biomass exceeding that in regions of sub-Saharan Africa that have been associated with the crop for centuries. There also remains considerable scope within its current distribution to increase bambara groundnut productivity by a clearer understanding of how factors such as the seasonal distribution of rainfall, daylength and range of temperatures influence the allocation of assimilates to pod yield. Ultimately, it is the expansion of production and consumption patterns for crops such as bambara groundnut both within and beyond their current distribution that

Table 8.1. Classification of suitability ranges for predicted biomass and pod yield of bambara groundnut.

	Very suitable	Suitable	Moderately suitable	Unsuitable
Biomass (kg ha⁻¹)	>8500	4500–8500	1500–4500	1000–1500
Pod yield (kg ha⁻¹)	>3000	1000–3000	300–1000	0–300

Note: kg ha⁻¹ refers to pod or biomass per crop and may not be restricted to one crop per year.

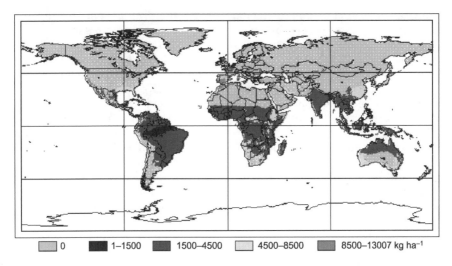

0 | 1–1500 | 1500–4500 | 4500–8500 | 8500–13007 kg ha⁻¹

Fig. 8.10. Predicted total biomass (kg ha^{-1}) for bambara groundnut across the world. (From Azam-Ali *et al.*, 2001b.)

will determine whether they become significant world crops or disappear.

Conclusions

One of the major concerns of sponsors is the possibility that effort is wasted on species of unknown potential in locations of unknown suitability. The analysis above demonstrates how a weather data generator and a dynamic crop simulation model can be linked in a GIS, to provide a spatial map of productivity on a global scale. In this way, many possible interactions can be explored that would be expensive and time consuming to determine in the field. Planners can then select promising locations, countries or regions that justify more detailed study, bringing in local factors and using data of higher resolution.

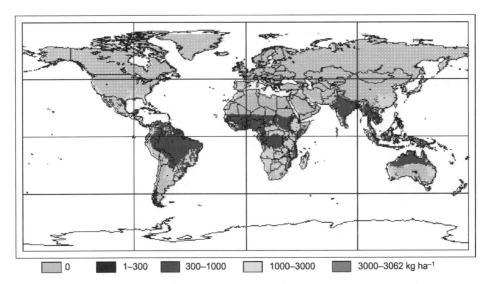

0 | 1–300 | 300–1000 | 1000–3000 | 3000–3062 kg ha⁻¹

Fig. 8.11. Predicted pod yield (kg ha^{-1}) for bambara groundnut across the world. (From Azam-Ali *et al.*, 2001b.)

The weather-generator and crop model can be enhanced as more data become available or the understanding of crop–weather relations increases. Perhaps the greatest constraint to verifying the yield maps at the scale of the country or region within a country is the limited information on both local climate and the physiological traits by which the model is parameterized. As information and expertise on bambara groundnut expand, however, landrace-specific information should provide estimates of likely productivity and best management practices for the cultivation of local bambara groundnut genotypes at particular locations. The development of a wider 'generic' methodology will provide comparative estimates of productivity for contrasting underutilized (and major) species. Moreover, the open and widespread access of the methodology and outputs to end-users and policy-makers will allow future developments to be demand-led by those most interested in the use of this information. In particular, comparison of the simulated yields of underutilized crops with existing yield maps for major crops will enable decision-makers to prioritize crop and cropping systems in terms of farmers' needs and national benefits. The potential effects of future change or variability in climate can similarly be considered.

Finally, although the approach described here is a first attempt to map the potential areas of cultivation of one underutilized food legume across the globe, it has wider implications in terms of food security and eliminating poverty. People eat a range of agricultural products that provide nutritional compounds including proteins, carbohydrates and lipids. The mapping strategy described here could lead to the production of 'nutritional maps' whereby the relative food values of different species could be assessed against production patterns for individual countries or regions. In this way, policy-makers can use evidence of the likely yield and nutritional value of different species to design food security strategies based on the most appropriate crops to grow within their region.

8.2.6 The diversity and enhancement of germplasm

Bambara groundnut has not been improved through coordinated breeding programmes and therefore different genotypes of this crop still exist as *landraces*. Zeven (1998) argued that only limited human selection might be enough to maintain a landrace distinct from other landraces. For bambara groundnut landraces, growers either save their own seed for the next season or buy various seed lots from the market. The mixing of seed before sowing, and natural selection during growth, both alter the genetic composition of the population. Landraces differ, therefore, from cultivars in many ways but the main distinction is that they are not uniform and stable in their characteristics from season to season. However, landraces are well buffered or homeostatic across a range of environments, a concept discussed by Simmonds and Smartt (1999) with regard to heterogeneous populations.

Morphological and agronomic traits have been widely used to assess the genetic variation of germplasm. For many species, especially under-researched crops such as bambara groundnut, it is still the only approach available to breeders and agronomists. One of bambara groundnut's attractive features is that, probably due to its wide geographic dispersion in Africa, there appears to be a wide range of genetic variability, which could give scope for selection. This is most obviously manifested as variation in seed size, testa colour and pattern, leaf, flower and pod numbers and traits conferring drought tolerance (Begemann, 1988; Collinson *et al.*, 1996; Squire *et al.*, 1997) and also photoperiod requirements for pod filling (Linnemann, 1994).

Among the more instructive findings is that, in bambara groundnut, vegetative and reproductive development are only weakly linked, whereas in most advanced ideotypes they are very tightly coupled. This is sometimes apparent at the scale of the whole stand, as when a heavy total dry mass is produced but little yield is obtained because of unsuitable photoperiod. It is more marked among individuals, such that, although most plants capture resources and produce a substantial biomass, relatively few flower and produce seed yield. For these diverse landraces, therefore, it might be necessary to assess resource capture and seed yield at an individual plant level, rather than at a canopy-level through bulked values of traits as normally applied to uniform crop stands (Squire *et al.*, 1997). This feature of bambara groundnut landraces has general implications for the ecophysiology of variable crop species, and is examined later in this chapter.

Molecular characterization

Experience in the use of molecular techniques in other crop species can, in principle, be applied to an underused crop such as bambara groundnut. Karp and Edwards (1997) provide a summary of methods. The simplest first approach is to use well tried techniques for measuring the variation in DNA sequences within and between populations (Tao *et al.*, 1993; Sharma *et al.*, 1995; He and Prakash, 1997). This might be followed by more detailed and comprehensive studies to identify genetic loci associated with the expression of agronomic traits, and eventually for marker-assisted selection (Lee *et al.*, 1996; Mian *et al.*, 1996; Nguyen *et al.*, 1997).

Some of these methods were accordingly applied to selected landraces of bambara groundnut. The first approach used AFLP (amplified fragment length polymorphism) and RAPD (random amplified polymorphic DNA) markers to evaluate genetic diversity (Massawe *et al.*, 1998; Massawe, 2000; Massawe *et al.*, 2000). RAPD and AFLP markers, variously, revealed high levels of DNA polymorphism generally, less variation within than between landraces, the scope for location or landrace specific markers for identification purposes, and a potential for association of AFLP markers and agronomic traits.

These attempts to define genetic diversity revealed what a nebulous concept the term 'landrace' is in many circumstances. Farmers either save seeds from the previous season or buy them from the market.

Scientists also usually buy seeds from the market or from individual growers, so every time experiments are conducted, different seed material is used. Often there is only cursory evidence for the maternal history of a so-called landrace. For example, it is our own experience that seeds bought in Gaborone market in Botswana were sold by individual entrepreneurs who had travelled from Zimbabwe specifically for this purpose. If this practice is commonplace across Africa, then local identity will generally be very difficult to establish: seed being distributed widely, mixed before sowing and acted on by local environment. To provide some basis for authenticity, landraces that are known to have been grown locally in different places in Africa are now being typed by molecular techniques. Seeds of these landraces will be multiplied under known conditions and a common seed lot will be used for future experiments across a number of seasons and locations.

Germplasm improvement

All the efforts described so far have attempted to evaluate various aspects of the nutritional, agroecological, genetic and marketing potential of existing bambara groundnut landraces. The choice of particular landraces for specific activities, e.g. nutritional composition, was inevitably governed by the availability of local germplasm and by the perceptions of the various scientists and participants involved in each activity. By definition, little was already known about the crop and the sequence of research activities therefore depended to a large extent on the availability of financial support and the specialist interests of participants at each institution rather than on an overarching strategy from the outset. Nevertheless, by 1999, the collective interest of research on bambara groundnut had generated sufficient critical mass and scientific evidence to justify further research to evaluate how the genetic potential of the existing germplasm might be improved by crop breeding and maximized by appropriate management.

In 2000, the EU Framework 5 programme agreed to support a new multidisciplinary project on bambara groundnut with partners in Africa and Europe. This project harnesses expertise in molecular biology, physiology, agronomy, modelling and agricultural extension and combines farmers and scientific knowledge within a common framework of analysis. Over 3 years, the project will provide the basis of a crop improvement programme that combines both conventional and novel methods of genetic evaluation and improvement. Crucially, a pre-requisite of the project is that it is based on the perceptions and preferences of the growers themselves and only uses scientific technologies to achieve objectives set by bambara groundnut farmers in each partner country. In this way, it is envisaged that the development of any bambara groundnut material, achieved either through the multiplication of existing landraces or the breeding of varieties, will be based on a conceptual ideotype identified by communities of farmers in each country rather than exclusively by scientists.

Among the guiding principles of the project are that conventional concepts of plant breeding, such as use of uniform cultivars and optimum growing conditions, need to be challenged. Landraces have a long history of successful adaptation to low-input agriculture: their survival is despite, rather than because of, plant breeders, agricultural scientists, extension agencies and sponsors. In breeding cultivars of bambara groundnut for difficult environments and poor farmers, selection must be conducted within the target environment and under the agronomic conditions of the farmers.

More specifically, the *potential* gain from a more uniform population in benign environments needs to be balanced against the *proven* advantages of non-uniformity in growth and development in marginal environments. Where rainfall is unpredictable, and where a more even distribution of food and income is desired, a highly variable population might have a number of disadvantages. The strategy behind a bambara groundnut breeding programme should therefore aim to balance the high-yielding

potential of individual plants with the yield stability provided by 'composite' landraces. The development of more uniform populations of plants allows the subsequent deployment of 'ideal blends', composed of a number of lines properly characterized for a set of agronomic traits.

Developments in molecular genetics offer new opportunities for trait-marker analysis, where certain phenotypic or morphological features can be associated or correlated with plant performance. The first step will be to identify bambara groundnut ideotypes for local conditions in Botswana, Namibia and Swaziland using a farmers survey. Since the start of the project, each African partner institution has initiated an extensive survey of local bambara groundnut farmers. The second main aim is to characterize the genetic and agronomic performance of a range of bambara groundnut landraces from Botswana, Namibia and Swaziland in controlled and on-farm environments. Field studies will be complemented by detailed physiological experiments in controlled environments to elucidate the specific responses of contrasting bambara groundnut landraces and individual plants to quantified variations in soil moisture, solar radiation and temperature. Subsequent tasks will include using a robust bambara groundnut model to match suitable ideotypes to contrasting environments and end-users, evaluation of genetic diversity in bambara groundnut germplasm using simple, readily transferable molecular tools, an operational method of crossbreeding for intraspecific hybridization and a strategic breeding programme for the species.

Finally, working with landraces has a number of implications for the field agronomist. The most important of these is probably the question of how to deal with biologically variable germplasm within physically and biologically variable environments. The convention of seeking to describe the average of the population or treatment level or the use of statistical techniques designed for crop varieties in relatively uniform plots should be reconsidered (see section 8.3).

8.2.7 Can experience of this crop be applied to other underutilized crops?

The question can now be asked as to whether the collective experience with bambara groundnut can be combined into a general methodological framework for examining the potential of any underused species. Here we explore the possible structure of a wider methodology and identify activities that are necessary for such a programme to be initiated.

Any general framework needs to be robust enough to integrate information from growers, on-station reports, local surveys and published literature. By combining existing information within a common format, a methodological framework should be able to provide growers and researchers with the most up-to-date information available on their species. For each location, it should also be possible to provide comparative information on other underutilized crops and major species.

Which crop? There is no easy answer to the question of which crops most deserve research attention. First, unlike more established species, evidence for the potential of underutilized crops is at best fragmentary and at worst contradictory. *Underutilized* crops are also *under-researched* crops and there are rarely more than a few reliable experimental studies in the literature for any underutilized crop. Evidence that does exist is often in the form of research station reports, copies of which may be difficult to obtain and contain results that may be difficult to collate with other evidence. Much of the information does not appear as published literature of any form but exists as anecdotal evidence, often uncorroborated by any conventional experimental research. This evidence may often have been obtained from informal surveys by organizations or individuals who themselves may have a vested interest in promoting a particular crop or crops. Selecting species to be included within a generic methodology will depend on the advocacy and co-operation of individuals and institutions involved in research or production of a particular species.

What research strategy? In practice, there are limitations to following a conventional approach. First, the infrastructure and financial support for research on underutilized crops is usually trivial in comparison with that devoted to more favoured species. Where funds are limited, it is imperative that the work that is done does not duplicate similar research being done elsewhere. Second, there is a limited amount of expertise that exists on any particular underutilized crop. It is important, therefore, that researchers take a multidisciplinary approach to maximize their contribution to knowledge on a particular crop rather than produce highly detailed work on an extremely narrow aspect of the crop without considering the wider implications of their work. Finally, there simply isn't the time to take a conventional approach. Most of the major crops of the world and their management requirements have been developed over many generations by a multitude of researchers, farmers, and commercial organizations. Neither the sponsors of similar research on largely untested underutilized species nor the end-users of this research, i.e. the growers and their families, can afford to wait years for the results of such research to be made available for practical management of field crops. Therefore, we require a methodology that combines as much of the information as possible on a crop so that we can identify both what we know and what we don't know about a particular crop in the minimum possible time.

How can results be disseminated? Unlike the major crop species, underutilized crops neither have an industrial nor an academic constituency and it is likely that any beneficiaries from research and extension activities will be the growers themselves. In such circumstances, the conventional route of publishing papers and presenting results at scientific meetings may not be the only or the best means of effectively disseminating findings. Methods must therefore be established that can convey research findings quickly and directly to the end-user and, wherever possible, obtain feedback on recommendations from growers back to the researchers.

The Internet offers possibilities of speed, directness and access to information that were unimaginable several years ago. Any attempts to widen and further personalize access to the internet for such purposes must be beneficial to those with interests in underused crops. Nevertheless, international workshops of all interested parties are highly valuable for establishing personal contacts, and are a relatively inexpensive item in the world of pan-global agricultural research.

A methodological framework

Any blueprint for developing an underutilized species must include a number of common factors. First, the users, especially the farmers and the cooks, must be part of the team. Second, the aims of the enterprise should be clearly defined, both from their point of view, and from that of those concerned at larger scales with the farming system, the long-term use of the land and the economics of a country. Third, the blueprint must include the means to combine a range of disciplines, each of which should be subservient to the collective aim. Fourth, the approach should take full advantage of existing methodologies, which are now advanced in most disciplines; and in doing so should include the latest global, technological developments, such as computer tools, crop simulation, socio-economic models, GIS and molecular biotechnology. This is not to support the indiscriminate use of global technology in subsistence agriculture, but to advocate the need to consider all opportunities, contentious or not, in furthering the aim. However difficult it may be to convince sponsors and molecular technologists, the development and application of novel technologies is at least as relevant in subsistence tropical agriculture as in commercial applications.

From experience with bambara groundnut, one of the more time-consuming and expensive of the activities is measuring the physiological traits needed to estimate potential yield in different environments. Even for the major crops, agronomic experiments are often limited by infrastructure and funding, and so are usually conducted for a limited number of sites and seasons. For underused species, an extensive experimental programme at many sites is simply unrealistic. More economical, and effective, is an integrated set of measurements in controlled environments and in the field, specifically aimed at measuring the set of traits. Incorporating these into a model of some form should enable a preliminary mapping exercise, as demonstrated above for bambara groundnut. Given this, it is possible to assess which crops are most likely to justify further research, where that research should be located for maximum returns and in which disciplines effort should be concentrated. In this way, the model will contribute to a methodology that combines published and indigenous knowledge of marketing, processing, agronomy and physiology with soils, climate and germplasm.

A plan of the methodological framework is shown in Fig. 8.12. The following are essential tasks.

1. Define the present distributions and role of the species in the cropping system, and as far as possible, reasons for its decline or limited use.
2. Determine that users have interest in the crop and its improvement.
3. Establish an information system for contacts and institutions involved with research on the crop.
4. Assess the nutritional, biochemical, processing and economic value of the products.
5. Measure the main traits in light-limited and water-limited growth, development and yield.
6. Define the agroecological yield potential and distribution, based on knowledge of soils and climate, using a mechanistic crop model.
7. On the basis of tasks 1–6, construct a plan of research for the development of the crop, to include coordination, logistics, timescale.
8. By experimentation and repeated referral back to the users, identify constraints to potential yield and stability of yield increases and appropriate management practices to overcome the constraints.
9. Identify the most appropriate genetic technology for assessing diversity of the

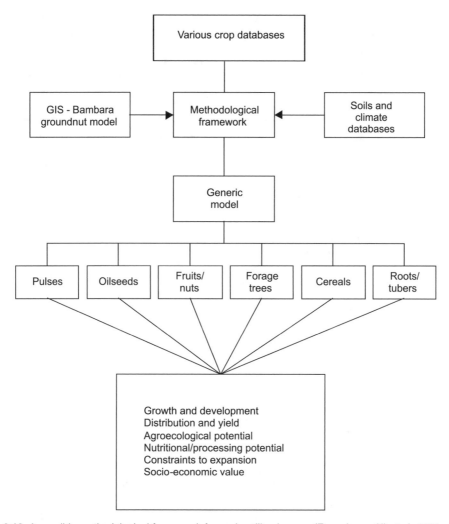

Fig. 8.12. A possible methodological framework for underutilized crops. (From Azam-Ali *et al.*, 2001a.)

crop, and implement a programme of genetic improvement that aims to match the variable germplasm with the variable environments of subsistence agriculture.

Progress can be made, and at relatively low cost compared to the spend on many topics of global agriculture. Piecemeal, *ad hoc* and duplicative research on different crops is unlikely to obtain support and ill-serves those who seek an improvement in the genetics and agronomy of underutilized species – the growers themselves. Finally, from our experience, it is clear that, as with all plant sciences, the greatest progress in agricultural biodiversity will be made through a multidisciplinary and multinational approach. Any narrow perspective based only on molecular reductionism, qualitative social sciences or empirical field trials is unlikely to be successful.

8.3 The status and contribution of agronomy

The experience with bambara groundnut showed that existing, fragmentary knowledge of a crop species combined with a basic understanding in ecophysiology was able to rapidly generate substantial new information that was valuable on both local and global scales. It raised the appreciation of crop–weather relations in this species to a level not far below that of the main tropical, leguminous crops. The opportunity for diversity in dryland cropping systems was thereby enhanced.

In this final section, we broaden the argument to consider some of the important matters facing tropical agriculture, specifically the sustainability of land usage, whether new biotechnology will make tropical agriculture better or worse and the matching of plant genotype to the properties of the desired production system. We reconsider the role of agronomy as an intermediary, among a range of disciplines, linking fundamental science to practice.

8.3.1 Resilience

Sustainability is a concept relating to human activity. An agricultural system is sustainable if we can continue exploiting it to produce food and other products without substantial and continued decline in output to the point where it no longer serves its functions. The foraging and exploitation activities of humans now so strongly affect all parts of the planet that any consideration of the sustainability of terrestrial ecosystems that is independent of agriculture does not hold. Clearly, agroecosystems will only sustain anything like the present global human population if they continue to produce the yield of the main staple crops and other products at something about the present rate per head of population. There might be variation in output – a surplus here, a deficit there – but there is not much slack left in the system.

How this continued output is best achieved is still the core of debate. Conway (1997) advocates a way forward through a combination of biotechnology and ecological science – deploying the most advanced biotechnology, which includes genetically modified (GM) crops, within a production system guided by sound ecological principles. This is a strong and compelling argument, but uncertainties arise over which biotechnology is admissible (see Section 8.3.2) and which ecological design is appropriate. One of Conway's arguments is that humans should, for their production systems, take the example of diverse, perennial 'natural' ecosystems. The argument relies on the notions that the greater the diversity of plants in a system, the greater its productivity, stability, resilience to disturbance and, by implication, its sustainability. However, the relations between diversity and these other descriptors are not definite or certain. Wood (1998) presents a critique of Conway (1997) and argues that the perennial, diverse model is not the most appropriate and should not be the most general. Wood argues that the appropriate ecological model for agriculture is early succession, the one most prevalent at present, where the species have rapid growth and intense competitive ability. Frequent disturbance is essential to allow the fast bulking rates of our annual crops' reproductive and storage organs. Investment in woody structures, as in later succession, diverts assimilate away from these nutritious organs.

Wood (1998) goes on to argue against the assumption that greater diversity causes greater stability and productivity. There are many productive single-species stands, and many, stable low-species systems. Some highly diverse ecosystems are very unstable in the face of human interference. Moreover, some farmers might establish diverse plots for a range of products near at hand, rather than for reasons of ecological stability. There are even good grounds for designing systems that are inherently unstable, in that they should be responsive to inputs. Stott (1998) further cautions against wider, uncritical assertions that stability is a desirable property and that ecological change is in some way bad for production systems. A descriptor such as resilience – the capacity of a system, following disturbance, to move back near its pre-disturbed state – is more apt when discussing agroecosystems. Even resilience is difficult to quantify, however, and we return to this discussion later in the chapter.

Stability of the soil

No single ecological model will be the right one for all circumstances. All agroecosystems must be both adaptable and diverse to a degree. Even those that are disturbed annually or several times a year are rarely stands of a single species: the crop coexists with weeds, and plant diversity is increased over a run of years by crop rotation, and by the introduction of new cultivars and species. However, the principal factor in maintaining stability is maintaining the state of the soil. If soil is washed or blown away faster than it is created by the binding of particles through microbial action, then no agroecosystem can continue to maintain or increase yield.

Syers (1998) summarizes the factors leading to degradation of soil and gives examples of how soils differ markedly in the rate at which productivity declines following loss of soil. Depletion of nutrients occurs when a crop is removed, in leaching and runoff (particularly N and K), erosion (particularly important for P) and in gaseous loss of N. Syers cites extracts from the study by Stoorvogel and Smaling (1992), whose nutrient balances, estimated for countries in sub-Saharan Africa, showed the inputs were invariably less than the outputs. Organic matter has a prime role both as a substrate for microbial communities and as a reserve of nutrients, particularly N, P and S.

There are very serious practical and scientific problems in managing soil to increase organic matter and realize its concomitant benefits. In most parts of the tropics, above-ground crop residues are needed for other purposes, such as feeding cattle. The quality and quantity of animal manure is insufficient by itself to return all that is removed. All sources must be used, including local and imported inorganic fertilizer. Syers (1998) also draws attention to the mas-

sive nutrient drain from some tropical countries in exported cash crops. A systematic transcontinental return of nutrients is surely feasible.

At a practical level, agronomy and related sciences can assess the state of a production system and indicate ways to improve it. Local action must be set in a broader perspective, as considered for rice by Greenland (1997). The first requirement is a simple nutrient balance at field or farm level (Guiking *et al.*, 1994). Agronomy can make defined contributions by calculating the present and potential gain and offtake of carbon and nutrients, given information on the available solar radiation and rainfall. It can then devise the most appropriate combinations of genotype (Chapter 4), timing (Chapter 5) and configuration (Chapter 6) from among the available crop types. For instance, it is quite feasible to estimate the number and type of crop species or cultivars (usually about three for any system) that result in the interception of >95% of solar radiation or rainwater.

There still remain many unknowns and uncertainties which research has the capacity to resolve. As Syers (1998) indicated, there is little information on threshold values of organic matter or other soil qualities below which deleterious effects occur. Further opportunities might come from understanding how the physical and chemical nature of plant residues, particularly the roots and their exudates, alter soil structure and function. There is also a need to define theoretical and experimental criteria for resilience of soil following perturbation. One of the main problems facing soil science is to go beyond the bulk measurements that are still used to describe the properties of soil. For instance, soil microbes absorb nutrients from breakdown of organic matter, and – since microbes turn over faster than organic matter itself – they make the nutrients more readily available to plants. However, there is increasing evidence that the total microbial biomass is not so important in determining microbial functioning: rather, it is the functional groups that are active in specific circumstances, and which might themselves be manipulable independently of the total biomass. The science of the soil at its finest scales still has major contributions to make to tropical agro-ecosystems.

8.3.2 Genetic technology and GM crops

All forms of plant breeding result in modification of the genome. Some techniques in 'conventional' genetics and breeding cause the combining of genes and traits in ways that would not be possible by cross-pollination. However, the particular terms 'genetic modification' or 'genetically modified' (GM) refer to instances where sequences of DNA are removed from one organism, are modified, augmented maybe, and then introduced and integrated with the DNA of another organism. Generally, the two organisms are of sexually incompatible plant species or even members of widely different taxonomic groups, such as higher plants and bacteria. The resulting GM or transgenic plants have altered biosynthetic pathways that confer different properties, such as resistance to a specific herbicide, or enhanced properties such as a greater fraction of a certain type of oil in the fruit. GM varieties have been widely grown in North America for several years, and are increasingly being developed and tested elsewhere, especially China. Among the first to be grown on a large scale were oilseed rape (*Brassica napus*) and maize varieties that are tolerant to broad-spectrum herbicides used to control a wide range of weed species, and cotton and maize that produce insecticidal toxins from inserted bacterial DNA. As background, Bruce and Bruce (1998) and Conway (1997, Chapter 8) provide examples and discussion for the lay reader, both of the technologies and of the ethical and other arguments surrounding their application to agriculture.

The fact that GM technology is possible has raised furious arguments, based on both practical and ethical issues. A range of opinions exists between two polarized stances. One is that the production and supply of food is generally limited by technology, that current technology is inadequate by itself to bring about the necessary improvements, and that only new technology can remedy the shortages. The other is that food is limited by unequal distribution, social inequality, civil strife, wars and natural catastrophe, and that existing technology would suffice if it had the chance, so nothing new is needed. Such stances and arguments would be unique to GM technology if it had potential consequences for good or bad that were intrinsically different from those of conventional breeding. There are certainly ethical arguments around transgenesis, but from a purely practical, agronomic stance, it is difficult to see that GM technology is inherently different from other agricultural technology, not least because conventionally bred varieties or the introduction of new crops from other countries have already caused, and will continue to cause, massive changes in food production and land use, and disruption to the pre-existing ecosystems, which have good or bad connotations depending on your viewpoint. Similarly, there should be no particular reason why GM crops should cause differential erosion of genetic resources or diminution of the gene pool, or more rapid loss of indigenous knowledge, than has been associated with non-GM crops.

There are of course instances where the tropics might be specifically disadvantaged by biotechnology. Modern genetic technologies might influence the economy of the tropics if cheaper substitutes for imported crops (e.g. certain oils) were able to be grown in temperate lands; or if trialling were carried out in tropical countries which had less stringent standards for assessing risks to health and to the environment, and some damage was caused as a result. A reasoned assessment at this stage of the technology's development is that risks of environmental harm are probably no greater than those associated with conventionally bred plant varieties, but that the benefits will not necessarily accrue more rapidly or more surely. Genetic modification is perhaps best considered dispassionately as a form of technology that has the potential to give great benefit to tropical crop ecology provided it can be introduced with due care and with the full knowledge and consent of the farmers and consumers.

Irrespective of its potential advantages, there remains the issue of widespread public acceptance of new genetic technologies. With this in mind, it is pertinent to ask: who funds research and development in GM crops; and for whom are the benefits of these technologies designed? Perhaps an unequivocal commitment from sponsors and researchers to GM technologies for poverty alleviation and food security rather than purely commercial or cosmetic benefits would encourage more support from the lay public and critics.

For agronomic science, the main issues to be resolved are therefore much the same as those to do with any technological change. Experience has shown that whenever technology is introduced, something happens that is not anticipated. This is simply because we do not understand the complexity of the biological interactions. If the events are considered damaging to health or environment, there may be a 'reactive' course taken to repair or ameliorate the situation. However, there are arguments in favour of a much more proactive or precautionary approach to new technology. The 'tiered' testing of GM varieties first in the laboratory, then in glasshouses, then in contained environments in the field has been applied in some instances and is an example of a precautionary approach. But at some point, a decision has to be made about field releases. For reasons just discussed, absolute safety or absence of effect can never be guaranteed. The way forward is to reach some balance between reactive and precautionary tactics, and the best way to facilitate this is by gaining basic knowledge of the production system and its links to and from the wider environment. Again, agronomic science can contribute by providing hard data on plant production and yields in relation to climate and soil; by quantifying the important physical or biotic constraints in any situation – so as to substantiate or refute claims that this or that genetic change will have such and such an effect; and (better still) by defining in the first place the type of product or process that is required to

enhance or diversify yield and make the system more resilient. Agronomy also has an important role in a wider biological, economic and social assessment of the hazards and benefits of any new technology.

Herbicide resistance in weeds and (GM) crops

Herbicide resistance provides a developing case study of the issues arising from the introduction of a GM crop. Several important species have been genetically transformed so as to possess resistance to specific, broad-spectrum herbicides. When sprayed with the herbicide, the crop remains unharmed but the weeds are likely to be killed. The potential benefits are that the farmer can be more flexible, spray to treat an emerging weed problem rather than spray in advance of a potential problem, and use fewer types of herbicides and perhaps less total herbicide, and that the broad-spectrum chemicals used, glyphosate and glufosinate, are denatured more quickly than many widely used, persistent herbicides. Potential disadvantages are that control of weeds will become so efficient that their function as diversifiers (for example, as harbourers of natural predators of pests) will dwindle, and as a result the biological diversity of agricultural ecosystems will further decrease.

Further potential drawbacks are that resistance to these herbicides will develop by natural selection in the weed flora, and will in specific instances be transferred to weedy relatives of the crop or to feral populations that are themselves weeds in a subsequent crop. More generally, resistance to active chemical ingredients in herbicides is not a trait peculiar to GM crops. Resistance must be present at very low frequency (e.g. 1 in a million individuals) in many weed species. Such individuals have a greater chance of surviving and reproducing when a herbicide is repeatedly used so that resistant types increase in frequency. Repeated use of the herbicide allows the resistant types to become dominant in the population of the species. Resistance has occurred in many weed taxa and is still occurring in many parts of the world (Moss and Rubin, 1993).

Again, there are ethical and practical points to do with transfer of herbicide resistance to weeds or wild relatives of crops. Hybridization has always occurred among certain crops and weeds, so there is nothing new in principle about this occurring with GM crops. Ethical issues include the potential for further mixing of genetic information among taxonomic strata that are not normally compatible (see Bruce and Bruce (1998) for discussion). The practical issues concern effective management and are the same whether herbicide resistance has been introduced through a GM crop or by selection operating on natural resistance.

Whichever way herbicide resistance enters a weed population, integrated management is seen as a means to counter it. The change to herbicide resistance in weeds can therefore be reversed by integrated approaches that use mechanical weed control, fallowing and the use of more competitive cultivars (Morgan, 1989). Moreover, the existence of a seedbank can slow the development of resistance; specifically, if non-resistant types emerge, flower and cross with resistant types, the resulting population will evolve more slowly to resistance.

8.3.3 An ecological synthesis

The main issues in ecology and agriculture are converging. Managed vegetation will increasingly have a range of purposes: provision of food or feed in the right quantity and quality, maintenance of soil resilience, habitat for other organisms, contribution to landscape. If the great advances in both molecular and ecological knowledge are to be integrated for the betterment of food production and wider diversity, how should this be done? Change can be brought about through the individual, most obviously by introducing one plant genotype or disadvantaging another. The effect is measured at the scale of the vegetation, the patch, subfield or field, in terms of, say, how much offtake and how much biological diversity can be supported in the same piece of land.

Answers are therefore required to two complementary types of question: (i) What is the potential effect of a group of plant genotypes, of defined physiological traits, on the properties of the vegetation? (ii) What plant genotypes and other organisms can co-exist within a piece of vegetation that is subject to defined limits, both environmental and of management?

Agronomy needs to consider the extent to which it can answer these questions. It could be argued that the reductionist philosophy prevalent in physiological science in the 1950s and 1960s delayed the application of physiological knowledge to issues of crop production. The causes were variously, ignorance of whole-organism biology, the institutionalized separation of laboratory and field scientists, and prevailing fashion. An understanding of the effect of plant process on yield can be made without reference to the molecular level, and indeed much emergent behaviour occurs in the whole plant and plant stand that is a direct result of context, and could not be predicted from fine-scale knowledge.

Looking back, it can also be argued that the main scientific advances that led to developments in agronomy in the previous 30 years did indeed come from integration of knowledge at the scale of the plant and crop stand and not from cellular or molecular physiology. From the 1960s onwards, agronomy learnt most from the systematic developments applied by environmental physics to the concepts of growth analysis introduced by Blackman (1919) and others (Evans, 1972). As part of the evolution of the disciplines, the complexities of environmental physics and crop physiology have been simplified and reduced over time, as in the example of flux-based growth analysis given by Goudriaan and Monteith (1990). Moreover, the thermal and photoperiodic responses, that were confirmed and quantified at the stand-scale by crop physiology, have been widely and successfully used as a standard for comparison of tropical species and for characterizing germplasm (Erskine et al., 1990; Summerfield et al., 1991).

It should now be considered whether

these concepts and tools alone can inform the more complicated matter of variable genotypes and variable microenvironments. Many of the advances brought about by applying environmental physics to crop physiology were possible because the plant stands being considered were uniform enough to be simplified by bulk characteristics such as the conversion factors for radiation and water (Chapters 2 and 3). Moreover, the physiological basis of most relations and models relies on there being little or no within-population variation, so that the mean of a trait-value at one developmental stage propagates downstream to determine the mean of another trait-value at a later development stage. The interception of radiation per unit field area during floral development, for instance, is taken to influence grain number per unit field area. But what if interception is to an extent uncoupled from reproduction? Experience with variable crops such as bambara groundnut, growing within highly variable environments, suggests this might be common, and calls for the need to question present methods of linking plant to environment.

The development of leaf number and flower number in two landraces of bambara groundnut illustrates the point (Fig. 8.13).

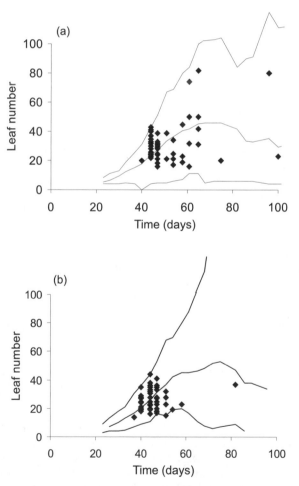

Fig. 8.13. Leaf production over time for samples of 50 plants from each of two landraces (a and b) of bambara groundnut in Sierra Leone, showing the progression of maximum, mean and minimum values of leaf number and (symbols) the leaf–time state at which each plant first flowered.

The mean, maximum and minimum leaf number among 50 plants grown in a field plot are shown together with the time and leaf number at which each individual first flowered. The two landraces displayed similar leafing states but different patterns of floral development superimposed on the leaf states. Flowering began slightly earlier in one landrace and in both began earlier on individuals that produced leaves more rapidly. Flowering was also more widely distributed over leaf–time states in one landrace compared with the other; and some individuals of this landrace did not flower, but still intercepted resource. The points showing time of first flower present the starting points for a distribution of flower states on time, much as the spread of emergence presented the starting points for a distribution of leaf states over time. The traces of leaf and flower states for all the individuals in a sampled population can be converted into maps, or 'phenoscapes', depicting the fraction of the population that appeared in various time–leaf–flower states. These configurations suggest that the progression of physiological trait-values through development is far from simple, and that stands might be better understood as a collection of individuals rather than a mean in highly variable conditions.

8.3.4 An individual-based approach to agronomy and ecology

A conceptual frame is needed which, with experimentation at the appropriate scale, will enable agronomy to link system and genotype. These two should be defined very carefully. The system is the aim of agronomy: the combination of vegetation and soil that satisfies the wants of the farmer and the wider society. The definition of the system must have a timescale and it must include those attributes beyond just the yield of a crop. The genotype is what agronomy introduces to the system, not in isolation but in the contexts of timing and configuration. Its definition must include variation, not just within species but also within crop varieties and populations. The physical link between

the two is the soil, which both affects and is affected by the performance of the genotype: the soil changes over time in response to genotype (and other factors), and in turn determines the genotypes that can be grown.

If the concept is to accommodate variable landraces, or any crops in poor conditions, it should deal in the physiology of individuals. Any individual is defined by both essential traits (that distinguish a sorghum from a cowpea, for instance) and the plasticity of those traits to conditions met as the plant grows. Plasticity has been widely observed in plant science but individual behaviour has seldom been considered. An example for crop plants of the latter is the study on maize by Edmeades and Daynard (1979). Neither plasticity nor characterization of individuals, however, is accommodated in typical conceptual frames for agriculture or ecology. Yet both could be, given present developments in ecological knowledge and computing power.

Referring to Fig. 8.13, each individual can be defined by a location in the many dimensional 'space' of its physiological traits. Two traits are indicated in that figure, but many others could be measured. The life cycle of each plant is then defined by its trajectory through the trait space, and the diversity of the plants in any system by the mesh of their trajectories (Squire *et al.*, 1997). A frame for linking genotype to system would be apparent, therefore, if the individuals, the mesh and their effect on the resource base could be measured and modelled. Indeed, many of the apparent differences between 'ecology' and 'agriculture' in their treatment of plant–environment relations would become irrelevant in any approach in which the individual plant is the unit of diversity.

One such approach is summarized in Fig. 8.14. Individuals are characterized by physiological traits describing their uptake and internal allocation of resource. The traits are quantified through *in situ* and *ex situ* methods, much as described for bambara groundnut, but with more attention to the individual. The individuals within a population, landrace, or species are then described as a distribution of individuals

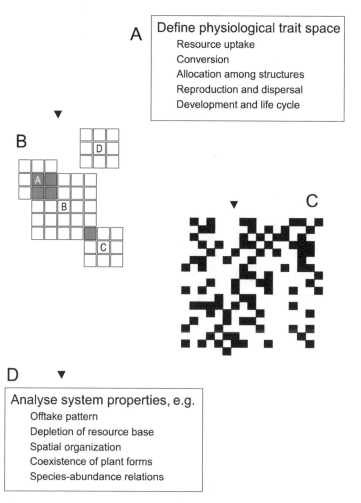

A   Define physiological trait space
     Resource uptake
     Conversion
     Allocation among structures
     Reproduction and dispersal
     Development and life cycle

B

C

D

Analyse system properties, e.g.
     Offtake pattern
     Depletion of resource base
     Spatial organization
     Coexistence of plant forms
     Species-abundance relations

Fig. 8.14. Steps in individual-based modelling, linking individual (genotype) to ecological system: (A) populations are characterized in terms of physiological traits; (B) individuals defined by these traits are configured on a substrate into which they extend and take up resource (e.g. plant B has a more extensive resource-capturing system than A, C, and D, and is competitively interacting with A and C but not D); (C) plants interact over time and give rise to emergent patterns in vegetation and substrate; (D) patterns are analysed to show effects on system-scale variables such as stand mass and nutrient content or biodiversity.

across this 'trait space'. A model is used to scale between trait and system, for example to predict system-scale effects from a change in trait. To do this, individuals are selected from the trait space according to the purpose in hand, and distributed in a defined configuration over a resource base, on which they grow and interact. The emerging system-scale properties are then analysed; such properties might be the distribution of bio-mass and nutrients among different crop plants and weeds over space and time, or features of the abundance relations of plant communities. These modelled, system-scale properties provide the hard, testable hypotheses that are needed to speed up the development and application of knowledge in this area.

Developments in field experimentation and statistical analysis will also contribute to the 'ecologizing' of agronomy. In order to confirm system-scale features arising from specific combinations of genotypes, some field experiments will require less uniform and rigid planting patterns than are typical of most agronomic experiments. Huxley (1999) considers the issues and consequences of variability in genotype and patchiness in environment.

Models and statistical analyses will benefit by being more open and accessible, especially if they are to be a medium for interaction between the agronomic, ecological and societal aspects of development in subsistence agriculture. Some possibilities are considered by Marshall et al. (1995). They argue that a model should be less of a 'black box', more open to interrogation both upstream (e.g. which variables affect a plant trait?) and downstream (e.g. what are the consequences of changing a plant trait on some later feature?). Models should also be able to take in data of different sorts, both quantitative and qualitative. Often, the existence of only qualitative information on a particular event or plant trait limits the predictive or explanatory power of ideas, but need not. Moreover, the prediction of a single value such as crop yield, with unknown uncertainty, is of limited practical use compared with a series of possible outcomes with associated probabilities. The belief networks described by Marshall et al. (1995) offer many desirable features, and could be readily adapted for use in studies of subsistence agriculture,

Concluding remarks

The agronomic decisions that farmers make have been guided by the accumulated store of their experiences and the traditions left by their predecessors, often over many centuries. Since the late 19th century, systematic research made increasing contributions by comparing the effectiveness of different practices to obtain the best possible yields at any location in response to the level and timing of inputs. Research did not always seek to understand the processes responsible for the improved yields because, in practical terms, understanding the underlying

mechanisms was of little benefit to the task in hand. The combination of traditional expertise, empirical evidence and improved varieties worked well, and can take much of the credit for the steady rise in crop yields that occurred up to the early part of the 20th century, particularly in temperate regions.

Knowledge and application progressed considerably during the last half-century. The factors determining evaporation and other micrometeorological exchanges between vegetation and the atmosphere became well understood, as did the broad physiological links between plant growth and environment. The processes of plant selection and breeding, and plant protection, were refined, and produced crop varieties and methods of growing them that strikingly surpassed anything that had gone before. As related, not all agricultural systems benefited from these developments. Moreover, variation in climate, and change in the soil consequent on heavy cropping, mean that existing high yields are by no means secure. The main challenges over the next 50 years will include maintaining production of the staple crops where their yields are already good, and improving the yield and quality of product in areas where these have hitherto remained poor.

The existing approaches to improving crop production might not be the most effective for achieving the tasks required in the next 50 years. The great variation in the soil of tropical subsistence environments, in the weather, and in the requirements of the people for a range of products, all have to be accommodated. The main aims should be production systems that have ecological resilience, while yielding a range of food and material products. Widening the range of crop ideotypes and matching variation in crop and soil, rather than smoothing it out, are likely to be essential.

Science has the means to initiate and support further major change in agriculture. Conceptual advances in ecology should lead to better means of assembling the right genotypes for a production system. The recent and increasingly rapid developments in genomics and gene manipulation will undoubtedly raise the rate at which plant breeding provides crop plants with new traits and new combinations of traits. The remarkable advances in understanding soil at the fine scale have yet to impinge on practice. Developments in modelling, field experimentation, and the use of statistics should all contribute. In this context, the role of agronomy as intermediary, as the focus of design, is ever more necessary.

References

Ahmed, S. and Rao, M.R. (1982) Performance of a maize/bean intercrop combination in the tropics: results of a multilocation study. *Field Crops Research* 5, 147–161.

Albrizio, R. (2001) Carbon dioxide and water vapour exchange rates of some Mediterranean field crops. PhD thesis, The University of Bari, Italy.

Allison, J.C.S. (1969) Effect of plant population on the production and distribution of dry matter in maize. *Annals of Applied Biology* 104, 357–365.

Altieri, M.A. (1995) *Agroecology. The Science of Sustainable Agriculture*. Westview Press, Boulder, Colorado.

Amthor, J.S. (1989) *Respiration and Crop Productivity*. Springer, New York.

Andrews, D.J. (1973) Effects of date of sowing on photosensitive Nigerian sorghums. *Experimental Agriculture* 9, 337–346.

Anonymous (1997) Proceedings of the International Symposium on Bambara Groundnut, 23–25 July 1996. University of Nottingham, UK.

Austin, R.B. (1983) Dry matter production and cropyield: genetic constraints. In: *The Nitrogen Requirements of Cereals*. MAFF/ADAS reference book 385. HMSO, London, pp. 41–50.

Austin, R.B. (1989) Genetic variation in photosynthesis. *Journal of Agricultural Science* 112, 287–294.

Austin, R.B., Bingham, J., Blackwell, R.D., Evans, L.T., Ford, M.A., Morgan, C.L. and Taylor, M. (1980) Genetic improvements in winter wheat yields since 1900 and associated physiological changes. *Journal of Agricultural Science* 94, 675–689.

Austin, R.B., Ford, M.A. and Morgan, C.L. (1989) Genetic improvement in the yield of winter wheat: a further evaluation. *Journal of Agricultural Science* 112, 295–301.

Ayad, W.G., Hodgkin, T., Jaradat, A. and Rao, V.R. (1997) *Molecular Genetic Techniques for Plant Genetic Resources*. International Plant Genetic Resources Institute, Rome.

Azam-Ali, S.N. (1984) Environmental and physiological control of transpiration by groundnut crops. *Agricultural and Forest Meteorology* 33, 129–140.

Azam-Ali, S.N. (1995) Assessing the efficiency of radiation use by intercrops. In: Sinoquet, H. and Cruz, P. (eds) *Ecophysiology of Tropical Intercropping*. INRA, Paris, pp. 305–318.

Azam-Ali, S.N., Gregory, P.J. and Monteith, J.L. (1984) Effects of planting density on water use and productivity of pearl millet (*Pennisetum typhoides*) grown on stored water. I. Growth of roots and shoots. *Experimental Agriculture* 20, 203–214.

Azam-Ali, S.N., Simmonds, L.P., Nageswara Rao, R.C. and Williams, J.H. (1989) Population, growth and water use of groundnut maintained on stored water. III. Dry matter, water use and light interception. *Experimental Agriculture* 25, 77–86.

Azam-Ali, S.N., Matthews, R.B., Williams, J.B. and Peacock, J.M. (1990) Light use, water uptake and performance of individual components of a sorghum/groundnut intercrop. *Experimental Agriculture* 26, 413–427.

Azam-Ali, S.N., Nageswara Rao, R.C., Craigon, J., Wadia, K.D.R. and Williams, J.H. (1993) A method for calculating the population/yield relations of groundnut (*Arachis hypogaea*) in semi-arid climates. *Journal of Agricultural Science* 121, 213–222.

Azam-Ali, S.N., Crout, N.M.J. and Bradley, R.G. (1994) *Perspectives in Modelling Resource Capture by Crops*. Nottingham University Press, Nottingham, UK.

Azam-Ali, S.N, Sesay, A., Karikari, S., Massawe, F.J., Aguilar-Manjarrez, J., Bannayan, M. and Hampson, K.J. (2001a) Assessing the potential of an underutilised crop – A case study using bambara groundnut. *Experimental Agriculture* 37, 479–518.

Azam-Ali, S.N., Aguilar-Manjarrez, J. and Bannayan-Avval, M. (2001b) *A Global Mapping System for Bambara Groundnut Production*. FAO Agricultural Information Management Series. No. 1. FAO, Rome, 48pp.

Babiker, A.M.A. (1989) Growth, dry matter yield of bambara groundnut (*Vigna subterranea*) and groundnut (*Arachis hypogaea*) under irrigated and drought conditions. MSc. thesis, University of Nottingham, UK.

Bachmann, K. (1994) Molecular markers in plant ecology. *New Phytologist* 126, 403–418.

Balkema-Boomstra, A.G. and Masterbroek, H.D. (1993) The grain yield of uniculm barley in two contrasting environments. *Euphytica* 66, 103–110.

Bationo, A., Christianson, C.B. and Mokwunye, U. (1989) Soil fertility management of the pearl millet producing soils of Sahelian West Africa: the Niger experience. Soil, crop and water management systems for rainfed agriculture in the Sudano-Sahelian zone. *Proceedings of the International Workshop*, 7–11 January 1987. ICRISAT, Sahelian Centre, Niamey, Niger, India, pp. 159–168.

Begemann, F. (1988) Ecogeographic differentiation of bambara groundnut (*Vigna subterranea*) in the collection of the International Institute of Tropical Agriculture (IITA). PhD thesis, Technical University Munich, Germany.

Bell, M.J., Muchow, R.C. and Wilson, G.L. (1987) The effect of plant population on peanuts (*Arachis hypogaea*) in a monsoonal tropical environment. *Field Crops Research* 17, 91–107.

Belsky, A.J. (1988) Regional influences on small-scale vegetational heterogeneity within grasslands in the Serengeti-National-Park, Tanzania. *Vegetatio* 74, 3–10.

Berchie, J.N. (1996) Light use and dry matter production of bambara groundnut landraces in relation to soil moisture. MSc. thesis, University of Nottingham, UK.

Blackman, V.H. (1919) The compound interest law and plant growth. *Annals of Botany* 33, 353–360.

Bleasdale, J.K.A. (1967) Systematic designs for spacing experiments. *Experimental Agriculture* 3, 73–86.

Blum, A. (1996) Yield potential and drought resistance: are they mutually exclusive? In: Reynolds, M.P., Rajaram, S. and McNab, A.

(eds) *Increasing Yield Potential in Wheat: breaking the barriers*. D.F. CIMMYT, Mexico.

Bonhomme, R. (1993) The solar radiation: characterization and distribution in the canopy. In: Varlet-Grancher, C., Bonhomme, R. and Sinoquet, H. (eds) *Crop Structure and Light Microclimate: characterization and applications*. INRA, Paris, pp. 17–28.

Brady, N.C. and Weil, R.R. (1999) *The Nature and Properties of Soils*, 12th edn. Prentice-Hall, London.

Brain, P. and Cousens, R. (1990) The effect of weed distribution on the prediction of yield loss. *Journal of Applied Ecology* 27, 735–742.

Briggs, D.J. and Courtney, F.M. (1989) *Agriculture and Environment. The Physical Geography of Temperate Agricultural Systems*. Longman Scientific, Essex, UK.

Brough, S.H. and Azam-Ali, S.N. (1992) The effect of soil moisture on the proximate composition of bambara groundnut (*Vigna subterranea* (L.) Verdc). *Journal of the Science of Food and Agriculture* 60, 197–204.

Brough, S.H., Taylor, A.J. and Azam-Ali, S.N. (1993) The potential of bambara groundnut (*Vigna subterranea*) in vegetable milk production and basic protein functionality systems. *Food Chemistry* 47, 277–283.

Bruce, B. and Bruce, A. (1998) *Engineering Genesis*. Earthscan, London.

Budyko, M.I. (1974) *Climate and Life*. Academic Press, New York.

Burkhill, I.H. (1935) A dictionary of the economic products of the Malay Peninsula. Ministry of Agriculture and Co-operatives, Kuala Lumpur. (Reprinted 1966).

Carr, M.K.V., Dale, M.O. and Stephens, W. (1987) Yield distribution in irrigated tea (*Camellia sinensis*) at two sites in Eastern Africa. *Experimental Agriculture* 23, 75–85.

Ceccarelli, S. and Grando, S. (1989) Efficiency of empirical selection under stress conditions in barley. *Journal of Genetics and Breeding* 43, 25–31.

Ceccarelli, S. and Grando, S. (1991) Environment of selection and type of germplasm in barley breeding for low-yielding conditions. *Euphytica* 57, 207–219.

Chandler, R.F. Jr (1969) Plant morphology and stand geometry in relation to nitrogen. In: Eastin, J.D., Haskins, F.A., Sullivan, C.Y. and Van Bavel, C.H.M. (eds) *Physiological Aspects of Crop Yield*. ASA, Madison, Wisconsin, pp. 265–285.

Charles-Edwards, D.A., Stutzel, H., Ferraris, R. and Beech, D.F. (1987) An analysis of spatial variation in the nitrogen content of leaves from different horizons within a canopy. *Annals of Botany* 60, 421–426.

Chikoye, D., Ekeleme, F. and Akobundu, I.O. (1997) Weed composition and population dynamics in intensified smallholder farms in West Africa. *Proceedings of the Brighton Crop Protection Conference – Weeds*. British Crop Protection Council, Farnham, UK, pp. 161–166.

Collins, W.W. and Hawtin, G.C. (1999) Conserving and using crop plant biodiversity in agro-ecosystems. In: Collins, W.W. and Qualset, C.O. (eds) *Biodiversity in Agroecosystems*. CRC Press, Boca Raton, Florida, pp. 267–282.

Collins, W.W. and Qualset, C.O. (1999) (eds) *Biodiversity in Agroecosystems*. CRC Press, Boca Raton, Florida, 334pp.

Collinson, S.T., Azam-Ali, S.N., Chavula, K.M. and Hodson, D.A. (1996) Growth, development and yield of bambara groundnut (*Vigna subterranea*) in response to soil moisture. *Journal of Agricultural Science* 126, 307–318.

Collinson, S.T., Clawson, E.J., Azam-Ali, S.N. and Black, C.R. (1997) Effect of soil moisture deficits on the water relations of bambara groundnut (*Vigna subterranea* L. Verdc.). *Journal of Experimental Botany* 48, 877–884.

Collinson, S.T., Berchie, J. and Azam-Ali, S.N. (1999) The effect of soil moisture on light interception and the conversion coefficient for three landraces of bambara groundnut (*Vigna subterranea*). *Journal of Agricultural Science* 133, 151–157.

Collinson, S.T., Sibuga, K.P., Tarimo, A.J.P. and Azam-Ali, S.N. (2000) Influence of sowing date on the growth and yield of bambara groundnut landraces in Tanzania. *Experimental Agriculture* 36, 1–13.

Connolly, J. (1986) On difficulties with replacement-series methodology in mixture models. *Journal of Applied Ecology* 23, 125–137.

Conway, G. (1997) *The Doubly Green Revolution: Food for All in the Twenty-first Century*. Penguin Books, London.

Cooper, P.J.M. (1979) The association between altitude, environmental variables, maize growth and yields in Kenya. *Journal of Agricultural Science* 93, 635–649.

Cooper, P.J.M. and Law, R. (1978) *Environmental and Physiological Studies of Maize, Part 3*, Vol. 1. Kenya Ministry of Agriculture, UK, Overseas Development Administration.

Cooper, P.J.M., Keatinge, J.D.H. and Hughes, G. (1983) Crop evapotranspiration – a technique for calculation of its components by field measurements. *Field Crops Research* 7, 299–312.

Cooper, P.J.M., Gregory, P.J., Keatinge, J.D.H. and Brown, S.C. (1987a) Effects of fertilizer, variety and location on barley production under rainfed conditions in Northern Syria. 2. Soil water dynamics and crop water use. *Field Crops Research* 16, 67–84.

Cooper, P.J.M., Gregory, P.J., Tully, D. and Harris, H.C. (1987b) Improving water use efficiency of annual crops in the rainfed farming systems of West Asia and North Africa. *Experimental Agriculture* 23, 113–158.

Corley, R.V.H. (1986) Yield potentials of plantation crops. *19th Colloquium, International Potash Institute*, Worblaufen-Bern, pp. 61–80.

Corley, R.H.V. (1977) Oil palm yield components and yield cycles. In: Earp, D.A. and Newall, W. (eds) *International Developments in Oil Palm*. Incorporated Society of Planters, Kuala Lumpur, pp. 116–129.

Corley, R.H.V., Hardon, J.J. and Tan, G.Y. (1971a) Analysis of growth of the oil palm (*Elaeis guineensis Jacq.*) 1. Estimating of growth parameters and application in breeding. *Euphytica* 20, 304–315.

Corley, R.H.V., Gray, B.S. and Ng, S.K. (1971b) Productivity of the oil palm (*Elaeis guineensis Jacq.*) in Malaysia. *Experimental Agriculture* 7, 129–136.

Coudert, M.J. (1984) Market openings in West Africa for cowpeas and bambara groundnuts. *International Trade Forum* 20, 14–29.

Cousens, R. (1985) A simple model relating yield loss to weed density. *Annals of Applied Biology* 107, 239–252.

Cousens, R. and Mortimer, M. (1995) *Dynamics of Weed Populations*. Cambridge University Press, Cambridge, UK.

Cox, T.S., Shroyer, J.P., Liu, B.-H., Sears, R.G. and Martin, T.J. (1988) Genetic improvement in agronomic traits of hard red winter wheat cultivars from 1919 to 1987. *Crop Science* 28, 756–60.

Cruz, P. and Soussana, J.F. (1997) Mixed crops. In: Lemaire, G. (ed.) *Diagnosis of the Nitrogen Status in Crops*. Springer, Berlin, pp. 131–144.

Curtis, D.L. (1968) The relation between yield and date of heading of Nigerian sorghums. *Experimental Agriculture* 4, 93–101.

Dalrymple, D.G. (1985) The development and adoption of high-yielding varieties of wheat and rice in developing countries. *American Journal of Agricultural Economics* 67, No. 5.

Dalrymple, D.G. (1986a) *Development and*

Spread of High-yielding Wheat Varieties in Developing Countries. USAID, Washington, DC.

Dalrymple, D.G. (1986b) *Development and Spread of High-yielding Rice Varieties in Developing Countries.* USAID, Washington, DC.

Darrah, L.L. and Penny, L.H. (1974) Altitude and environmental responses of entries in the 1970–71 East African maize variety trial. *East African Agriculture and Forestry Journal* 40, 77–88.

Davis, J.H.C. and Garcia, S. (1983) Competitive ability and growth habit of indeterminate beans and maize for intercropping. *Field Crops Research* 6, 59–75.

De Wit, C.T. (1958) *Transpiration and Crop Yields.* Wageningen Agricultural Research Reports 64.6. Institute of Biological and Chemical Research on Field Crops and Herbage, Wageningen, Netherlands.

Dingkuhn, M., Penning de Vries, F.W.T., De Datta, S.K. and van Laar, H.H. (1991) Concepts for a new plant type for direct seeded flooded tropical rice. In: *Direct Seeded Flooded Rice in the Tropics.* IRRI, Los Banos, Philippines.

Doku, E.V. (1969) Growth habit and pod production in bambara groundnut (*Voandzeia subterranea*). *Ghana Journal of Agricultural Science* 2, 91–95.

Doku, E.V. (1997) Problems and prospects for the improvement of bambara groundnut. *International Bambara Groundnut Symposium*, 23–25 July 1996, University of Nottingham, Nottingham, UK.

Donald, C.M. (1968) The breeding of crop ideotypes. *Euphytica* 17, 385–403.

Donald, C.M. and Hamblin, J. (1976) The convergent evolution of annual seed crops in agriculture. *Advances in Agronomy* 36, 97–143.

Doorenbos, J. and Kassam, A.H. (1986) *Yield Responses to Water.* FAO irrigation and drainage paper 33. FAO, Rome.

Doorenbos, J. and Pruitt, W.O. (1977) *Crop Water Requirements.* FAO irrigation and drainage paper 21. FAO, Rome.

Doyle, C.J. (1991) Mathematical models in weed management. *Crop Protection* 10, 432–444.

Duncan, W.G. (1971) Leaf angles, leaf area and canopy photosynthesis. *Crop Science* 11, 482–485.

Eberhart, S.A., Penny, L.H. and Harrison, M.N. (1973) Genotype by environment interactions in maize in eastern Africa. *East African Agriculture and Forestry Journal* 39, 61–71.

Edmeades, G.O. and Daynard, T.B. (1979) The development of plant-to-plant variability in maize at different planting densities. *Canadian Journal of Plant Science* 59, 561–576.

El Sharkawy, M.A. and Cadavid, L.F. (2000) Genetic variation within cassava germplasm in response to potassium. *Experimental Agriculture* 36, 323–334.

Erskine, W., Ellis, R.H., Summerfield, R.J., Roberts, E.H. and Hussain, A. (1990) Characterization of responses to temperature and photoperiod for time to flowering in a world lentil collection. *Theoretical and Applied Genetics* 80, 193–199.

Evans, G.C. (1972) *The Quantitative Analysis of Plant Growth.* Blackwell, Oxford.

Evans, L.T. (1993) *Crop Evolution, Adaptation and Yield.* Cambridge University Press, Cambridge, UK.

Evans, L.T. and Dunstone, R.L. (1970) Some physiological aspects of evolution in wheat. *Australian Journal of Biological Sciences* 23, 725–741.

Falconer, D.S. (1990) Selection in different environment: effects on environmental sensitivity (reaction norm) and on mean performance. *Genetics Research* 56, 56–70.

Farquhar, G.D. and von Caemmerer, S. (1982) *Modelling of Photosynthetic Response to Environmental Conditions.* Springer, Berlin.

Field, C. (1983) Allocating leaf nitrogen for the maximization of carbon gain: leaf age as a control on the allocation program. *Oecologia* 56, 341–347.

Finlay, K.W. and Wilkinson, G.N. (1963) The analysis of adaptation in a plant breeding programme. *Australian Journal of Agricultural Research* 14, 742–754.

Fischer, K.S. and Palmer, A.F.E. (1983) Maize. In: *Potential Productivity of Field Crops under Different Environments.* IRRI, Los Banos, Phillippines, pp. 155–180.

Fischer, R.A. (1979) Growth and water limitation to dryland wheat yields in Australia: a physiological framework. *Journal of the Australian Institute of Agricultural Science* 45, 83–94.

Fischer, R.A. (1981) Optimising the use of water and nitrogen through breeding of crops. *Plant and Soil* 58, 249–278.

Fischer, R.A. (1976) Wheat physiology at CIMMYT and raising the yield plateau. In: Reynolds, M.P., Rajaram, S. and McNab, A. (eds) *Increasing Yield Potential in Wheat: Breaking the Barriers.* D.F. CIMMYT, Mexico, pp. 195–202.

Fischer, R.A. and Wood, J.R. (1979) Drought resistance in spring wheat cultivars. III.

Yield associations with morpho-physiological traits. *Australian Journal of Agricultural Research* 30, 1001–1020.

Fofana, B., Koupeur, T., Jones, M.P. and Johnson, D.E. (1995) *The Development of Rice Varieties Competitive with Weeds.* Brighton Crop Protection Conference – Weeds 1995, pp. 187–192.

Food and Agriculture Organisation of the United Nations (1984) *FAO Fertilizers Yearbook.* FAO, Rome.

Food and Agriculture Organisation of the United Nations (1996) *Report on the State of the World's Plant Genetic Resources for Food and Agriculture.* FAO, Rome.

Food and Agriculture Organisation of the United Nations (1998) *FAO Production Yearbook,* Vol. 52. FAO, Rome.

Ford-Lloyd, B. and Painting, K. (1996) *Measuring Genetic Variation Using Molecular Markers.* International Plant Genetic Resources Institute, Unit 10.1.4. http://www.cgiar.org/ipgri.

Forster, B.P, Lee, M.A., Lundquist, U., Millam, S., Vamling, K. and Wilson, T.M.A. (1997) Genetic engineering of crop plants: from genome to gene. *Experimental Agriculture* 33, 15–33.

Foster, H.L. (1976) Yield response of oil palm to fertilisers in West Malaysia. 1. Yield response functions. *MARDI Research Bulletin* 4, 44–63.

Foster, H.L. and Goh, H.S. (1977) Yield response of oil palm to fertilisers in West Malaysia. II. Influence of soil and climatic factors. *MARDI Research Bulletin* 5, 6–22.

Frankel, O.H., Brown, A.H.D. and Burden, J.J. (1995) *The Conservation of Plant Biodiversity.* Cambridge University Press, Cambridge, UK.

Frost, H.M. (1995) *Striga hermonthica* surveys in Western Kenya. *Proceedings of the Brighton Crop Protection Conference – Weeds.* British Crop Protection Council, Farnham, UK, pp. 145–150.

Fussell, L.K, Serafina, P.G., Bationo, A. and Klaij, M.C. (1987) Management practices to increase yield and yield stability of pearl millet in Africa. *Proceedings of the International Pearl Millet Workshop,* 7–11 April 1976, ICRISAT Centre, India, pp. 255–266.

Gallagher, J.N. and Biscoe, P.V. (1978) Radiation absorption, growth and yield of cereals. *Journal of Agricultural Science* 91, 47–60.

Gallagher, J.N., Biscoe, P.V. and Scott, R.K. (1975) Barley and its environment. V. Stability of grain weight. *Journal of Applied Ecology* 12, 319–336.

Gardner, F.P. and Auma, E.O. (1989) Canopy structure, light interception, and yield market quality of peanut genotypes as influenced by planting pattern and planting date. *Field Crops Research* 20, 13–29.

Gerwitz, A. and Page, E.R. (1974) An empirical mathematical model to describe plant root systems. *Journal of Applied Ecology* 11, 773–782.

Gill, G.J. (1991) *Seasonality and Agriculture in the Developing World. A Problem of the Poor and Powerless.* Cambridge University Press, Cambridge, UK.

Giller, K.E. (2001) *Nitrogen Fixation in Tropical Cropping Systems,* 2nd edn. CAB International, Wallingford, UK.

Goel, N.S. (1988) Models of vegetation canopy reflectance and their use in estimation of biophysical parameters from reflectance data. *Remote Sensing Reviews* 4, 1–212.

Goldsworthy, P.R. (1970) The sources of assimilate for grain development in tall and short sorghum. *Journal of Agricultural Science* 74, 523–531.

Goldsworthy, P.R. and Colegrove, M. (1974) Growth and yield of highland maize in Mexico. *Journal of Agricultural Science* 83, 213–221.

Goudriaan, J. (1977) *Crop Micrometeorology: a Simulation Study.* Pudoc, Wageningen.

Goudriaan, J. and Monteith, J.L. (1990) A mathematical function for crop growth based on light interception and leaf area expansion. *Annals of Botany* 66, 695–701.

Greenland, D.J. (1995) Long-term cropping experiments in developing countries: the need, history and future. In: Leigh, R.A. and Johnson, A.E. (eds) *Long-term Experiments in Agricultural and Ecological Sciences.* CAB International, Wallingford, pp. 187–209.

Greenland, D.J. (1997) *The Sustainability of Rice Farming.* CAB International, Wallingford, UK.

Gregory, P.J., Shepherd, K.D. and Cooper, P.J. (1984) Effects of fertilizer on root growth and water use of barley in Northern Syria. *Journal of Agricultural Science* 103, 429–438.

Guiking, T.F.C., Jansen, D.M. and Fresco, L.O. (1994) The use of simplified nutrient balances at farm level to determine boundary conditions for sustainable production. In: Syers, J.K. and Rimmer, D.L. (eds) *Soil Science and Sustainable Land Management* in the Tropics. CAB International, Wallingford, pp. 248–257.

Hall, A.E., Thiaw, S. and Krieg, D.R. (1994) Consistency of genotypic ranking for carbon isotope discrimination by cowpea grown in tropical and subtropical zones. *Field Crops Research* 36, 125–131.

Hall, A.J., Whitfield, D.M. and Connor, D.J. (1990) Contribution of preanthesis assimilates to grain-filling in irrigated and water-stressed sunflower crops. II. Estimates from a carbon budget. *Field Crops Research* 24, 273–294.

Hamdi, Q.A., Harris, D. and Clark, J.A. (1987) Saturation deficit, canopy formation and function in *Sorghum bicolor* (L.). *Journal of Experimental Botany* 38, 1272–1283.

Hamel, C., Barrantes-Cartin, U., Furlan, V. and Smith, D.L. (1991) Endomycorrhizal fungi in nitrogen transfer from soybean to maize. *Plant Soil* 138, 33–40.

Hampson, K., Azam-Ali, S.H., Sesay, A., Mukwaya, S.M. and Azam-Ali, S.N. (2000) *Assessing opportunities for increased utilisation of bambara groundnut in Southern Africa.* Tropical Crops Research Unit, School of Biosciences, University of Nottingham. DFID CPHP Project R–7527. http://www.genres.de/bambara/r_7527.

Hanks, R.J., Keller, J., Rasmussen, V.P. and Wilson, G.D. (1976) Line source sprinkler for continuous variable irrigation crop production studies. *Soil Science Society of America Journal* 40, 426–429.

Hanson, H., Borlaug, N.E. and Anderson, R.G. (1982) *Wheat in the Third World.* Westview Press, Boulder, Colorado.

Harlan, J.R. (1971) Agricultural origins: centers and noncenters. *Science, Washington* 174, 46–474.

Harlan, J.R. and de Wet, J.M.J. (1972) A simplified classification of cultivated sorghum. *Crop Science* 12, 172–176.

Harland, H.V and Anthony, S. (1920) Development of barley kernels in normal and clipped spikes and the limitations of awnless and hooded varieties. *Journal of Agricultural Research* 19, 431–472.

Harper, J.L. (1977) *Population Biology of Plants.* Academic Press, London.

Harris, D., Natarajan, M. and Willey, R.W. (1987) Physiological basis for yield advantage in a sorghum/groundnut intercrop exposed to drought. 1. Dry matter production, yield and light interception. *Field Crops Research* 17, 259–272.

Haverkoort, A.S.J. and Mackerron, D.K.L. (2000)

Management of Nitrogen and Water in Potato Production. Wageningen, The Netherlands.

Hawkes, J.G. (1983) *Diversity of Crop Plants.* Harvard University Press.

Hawkins, R.C. and Cooper, P.J.M. (1981) Growth, development and grain yield of maize. *Experimental Agriculture* 17, 203–207.

Hay, R.K.M. and Kirby, E.J.M. (1991) Convergence and synchrony – a review of the coordination of development in wheat. *Australian Journal of Agricultural Research* 42, 661–700.

Hazell, P.B.R. and Ramasamy, C. (1991) *The Green Revolution Reconsidered: the Impact of High Yielding Rice Varieties in South India.* Johns Hopkins University Press, Baltimore.

He, G.H. and Prakash, C.S. (1997) Identification of polymorphic DNA markers in cultivated peanut (*Arachis hypogaea* L.). *Euphytica* 97, 143–149.

Hepper, F.N. (1963) Plants of the 1957–58 West Africa expedition. The bambara groundnut (*Voandzeia subterranea*) and kersting's groundnut (*Kerstingiella geocarpa*) wild in West Africa. *Kew Bulletin* 16, 395–407.

Hiebsch, C.K. and McCollum, R.E. (1987) Area × Time Equivalency Ratio. A method for evaluating the productivity of intercrops. *Agronomy Journal* 79, 15–22.

Hill, J. (1975) Genotype–environment interactions – a challenge for plant breeding. *Journal of Agricultural Science* 85, 477–493.

Hirose, T., Werger, M.J.A., Pons, T.L. and van Rheenen, W.A. (1988) Canopy structure and leaf nitrogen distribution in a stand of *Lysimachia vulgaria* L. as influenced by stand density. *Oecologia* 77, 145–150.

Holliday, R. (1960) Plant population and crop yield. *Nature* 186, 22.

Horst, W. (1995) Efficiency of soil-nutrient use in intercropping systems. In: Sinoquet, H. and Cruz, P. (eds) *Ecophysiology of Tropical Intercropping.* INRA, Paris, pp. 197–212.

Howard, S.B., Ong, C.K., Rao, M.R., Mathuva, M. and Black, C.R. (1995) The partitioning of light and water in Leucaena-maize agroforestry systems. In: Sinoquet, H. and Cruz, P. (eds) *Ecophysiology of Tropical Intercropping.* INRA, Paris, pp. 123–136.

Hsiao, T.C. (1982) The soil–plant–atmosphere continuum in relation to drought and crop production. In: *Drought Resistance in Crops with Emphasis on Rice.* IRRI, Los Banos, Philippines, pp. 39–52.

Huda, A.S.K., Sivakumar, M.V.K. and Rego, T.J. (1985) An analysis of agro-climatic data in relation to fertiliser use in dryland farming areas of India. *Fertiliser News* 30, 35–41.

Hughes, G. (1986) An analysis of the effects of variable emergence and harvest dates on solar radiation interception and loss by crop canopies. *Crop Research* 26, 41–48.

Hughes, G. (1996) Incorporating spatial pattern of harmful organisms into crop loss models. *Crop Protection* 15, 407–421.

Hughes, G. and Keatinge, J.D.H. (1983) Solar radiation interception, dry matter production and yield in pigeon pea (*Cajanus cajan* L. Millspaugh). *Field Crops Research* 6, 171–178.

Hughes, G., Keatinge, J.D.H. and Scott, S.P. (1981) Pigeon pea as a dry season crop in Trinidad, West Indies. II. Interception and utilisation of solar radiation. *Tropical Agriculture (Trinidad)* 58, 191–199.

Hurd, E.A. (1971) Can we breed for drought resistance? In: Larson, K.L. and Eastin, J.D. (eds) *Drought Injury and Resistance in Crops*. Special Publication No. 2. Crop Science Society of America, Madison, Wisconsin, pp. 77–88.

Hutchinson, J.B., Manning, H.L. and Farbrother, H.G. (1958) Crop water requirement of cotton. *Journal of Agricultural Science* 51, 177–188.

Huxley, P. (1999) Patchiness; living in the real world. In: *Tropical Agroforestry*. Blackwell Science, Oxford, pp. 241–246.

IBPGR/IITA/GTZ (1987) *Descriptors for Bambara Groundnut*. International Board for Plant Genetic Resources, Rome, Italy, 23 pp.

Innes, N.L. (1992a) The contribution from conventional plant breeding. *Proceedings of the Royal Society of Edinburgh* 99B, 1–10.

Innes, N.L. (1992b) Gene banks and their contribution to the breeding of disease resistant cultivars. *Euphytica* 63, 23-31.

IRRI (1989) *IRRI towards 2000 and beyond*. International Rice Research Institute, Los Banos, Philippines.

Jaetzold, T. and Kutsch, H. (1980) *Climatic data bank of Kenya*. Department of Cultural and Regional Geography, University of Trier, Germany.

Jaetzold, T. and Schmidt, H. (1982) *Farm Management Handbook of Kenya*, 3 vols. Ministry of Agriculture, Nairobi.

Jenny, H. (1980) *The Soil Resource*. Ecological Studies, Vol. 37. Springer-Verlag, New York.

Johnson, D.T. (1968) The bambara groundnut, a review. *Rhodesian Agricultural Journal* 65, 1–4.

Jones, E., Nyamudesa, P. and Busangavanye, T.

(1989) Rainfed cropping and water conservation and concentration on vertisols in the SE lowveld of Zimbabwe. In: *Proceedings of a Workshop on Vertisol Management in Africa, January 1989, Harare*. International Board of Soil Research and Management (IBSRAB), Bangkok, Thailand.

Karikari, S.K. (1996) The status of bambara groundnut genetic resources in Botswana. *SACCAR Newsletter* No 34, June, 1996.

Karikari, S.K., Chaba, O. and Molosiwa, B. (1999) Effects of intercropping bambara groundnut on pearl millet, sorghum and maize in Botswana. *African Crop Science Journal* 7, 143–152.

Karp, A. and Edwards, K.J. (1997) *Molecular Techniques in the Analysis of the Extent and Distribution of Genetic Diversity*. Report of an IPGRI Workshop, 9–11 October 1995, Rome, Italy, pp. 11–22.

Kassam, A.H. and Andrews, D.J. (1975) Effects of sowing date on growth, development and yield of photosensitive sorghum at Samaru, northern Nigeria. *Experimental Agriculture* 11, 227–240.

Kassam, A.H., Kowal, J.M. and Harkness, C. (1975) Water use and growth of groundnut at Samaru, northern Nigeria. *Tropical Agriculture* 52, 105–112.

Keating, B.A. and Carberry, P.S. (1993) Resource capture and use in intercropping – solar-radiation. *Field Crops Research* 34, 273-301.

Keating, B.A., Godwin, D.C. and Watiki, J.M. (1991) Optimising nitrogen inputs in response to climatic risk. In: Muchow, R.C. and Bellamy, J.A. (eds) *Climatic Risk in Crop Production: Models and Management for the Semiarid Tropics and Subtropics*. CAB International, Wallingford, pp. 329-358.

Khush, G.S. and Peng, S. (1996) In: Reynolds, M.P., Rajaram, S. and McNab, A. (eds) *Increasing Yield Potential in Wheat: Breaking the Barriers*. D.F. CIMMYT, Mexico, pp. 36–51.

Kingdon-Ward, F. (1952) Does wild tea exist? *Nature* 165, 297–299.

Kiniry, J.R., Jones, C.A., O'Toole, J.C., Blanchet, R., Cabelguenne, M. and Spanel, D.A. (1989) Radiation-use efficiency in biomass accumulation prior to grain-filling for five grain-crop species. *Field Crops Research* 20, 51–64.

Kocabas, Z., Craigon, J. and Azam-Ali, S.N. (1999) The germination response of bambara groundnut to temperature. *Seed Science and Technology* 27, 303-313.

Koide, R.T., Robichaux, R.H., Morse, S.R. and Smith, C.M. (1994) Plant water status, hydraulic resistance and capacitance. In: Pearcy, R.W., Ehleringer, J., Mooney, H.A. and Rundel, P.W. (eds) *Plant Physiological Ecology, Field Methods and Instrumentation*. Chapman and Hall, London, pp. 161–183.

Koppen, W. (1931) *Grundrisz der Klimakunde*. Walter de Gruyter, Berlin and Leipzig, 388pp.

Koppen, W. (1936) Das geographische System der Klimate, Berlin, Gebr.Borntrager, Vol. 1. Part C of Koppen-Geiger, Handbuch der Klimatologie, 44pp.

Kowal, J.M. and Kassam, A.H. (1973) Water use, energy balance and growth of maize at Samaru, northern Nigeria. *Agricultural Meteorology* 12, 391–406.

Kowal, J.M. and Knabe, D. (1972) *An Agroclimatological Atlas of the Northern States of Nigeria*. Ahmadu Bello University Press, Zaria.

Kropff, M.J. and van Laar, H.H. (eds) (1993) *Modelling Crop–Weed Interactions*. CAB International, Wallingford.

Lamb (1985) Rainfall in Subsaharan West Africa during 1941–83. *Zeitschrift fur Gletscherkunde und Glazialgeologie* 21, 131–139.

Lane, P., Galwey, N. and Alvey, N. (1987) *Genstat 5: an Introduction*. Clarendon Press, Oxford.

Lee, S.H., Bailey, M.A., Mian, M.A.R., Carter, T.E., Ashley, D.A., Hussey, R.S., Parrott, W.A. and Boerma, H.R. (1996) Molecular markers associated with soybean plant height, lodging and maturity across locations. *Crop Science* 36, 28–735.

Lemaire, G. and Gastal, F. (1997) N uptake and distribution in plant canopies. In: *Diagnosis of the Nitrogen Status in Crops*. Springer, Berlin, 239pp.

Lemaire, G., Onillon, B., Gosse, G., Chartier, M. and Allirand, J.M. (1991) Nitrogen distribution within a lucerne canopy during regrowth: relation with light distribution. *Annals of Botany* 68, 483–488.

Lemcoff, J.H. and Loomis, R.S. (1986) Nitrogen influences on yield determination in maize. *Crop Science* 26, 1017–1022.

Lemeur, R. and Blad, B.L. (1974) A critical review of light models for estimating the shortwave radiation regime of plant canopies. *Agricultural Meteorology*, 14, 255–286.

Linnemann, A.R. (1994) Photothermal regulation of phenological development in bambara

groundnut (*Vigna subterranea* L. Verdc.). PhD thesis Wageningen Agricultural University, The Netherlands.

Linnemann, A.R. and Azam-Ali, S.N. (1993) Bambara groundnut (*Vigna subterranea*). In: Williams, J.T. (ed.) *Underutilized Crops. Pulses and Vegetables*. Chapman and Hall, London, pp. 13–57.

Lockwood, J.G. (1974) *World Climatology. An Environmental Approach*. Edward Arnold, London, 330 pp.

Loomis, R.S. and Connor, D.J. (1992) *Crop Ecology: Productivity and Management in Agricultural Systems*. Cambridge University Press, Cambridge.

Ludlow, M.M. and Muchow, R.C. (1990) A critical evaluation of traits for improving crop yields in water limited environments. *Advances in Agronomy* 43, 107–153.

McBratney, A.B. (1992) On variation, uncertainty and informatics in environmental soil management. *Australian Journal of Soil Research* 30, 913-35.

McCance, R.A. and Widdowson, E. (1978) The composition of foods. In: Paul, A.A. and Southgate, D.A.T. (eds) *Fourth Revised and Extended Edition of MRC Special Report no. 297*. HMSO, London, 209pp.

McCree, K.J. and Silsbury, J.H. (1978) Growth and maintenance requirements of subterranean clover. *Crop Science* 19, 13–18.

Marini-Bettolo, G.B. (1987) Scientific research and the challenge of agriculture in the tropics. In: Marini-Bettolo, G.B. (ed.) *Towards a Second Green Revolution. From Chemical to New Biological Technologies in Agriculture in the Tropics*. Elsevier, Amsterdam, pp. 7–10.

Marshall, B. and Willey, R.W. (1983) Radiation interception and growth in an intercrop of pearl millet/groundnut. *Field Crops Research* 7, 141–160.

Marshall, B., Crawford, J.W. and McNicol, J. (1995) Handling qualitative and uncertain information. In: Haverkort, A.J. and MacKerron, D.K.L. (eds) *Potato Ecology and Modelling of Crops Under Conditions Limiting Growth*. Kluwer Academic Press, London, pp. 323-340.

Marshall, B., Crawford, J.W. and Porter, J.R. (1997) Variability and scaling: matching methods and phenomena. In: Van Gardingen, P.R., Foody, G.M. and Curran, P.J. (eds) *Scaling-up*. Cambridge University Press, Cambridge, UK, pp. 253–272.

Marshall, F.M., Black, C.R. and Ong, C.K. (1994) Heat balance measurements of transpiration in perennial pigeonpea–groundnut agroforestry systems. In: Monteith, J.L., Scott, R.K. and Unsworth, M.H. (eds) *Resource Capture by Crops*. Proceedings of 52nd University of Nottingham Easter School. Nottingham University Press, Loughborough, UK, pp. 426–429.

Massawe, F.J. (2000) Phenotypic and genetic diversity in bambara groundnut landraces. PhD thesis, University of Nottingham, UK.

Massawe, F.J., Roberts, J.A. and Azam-Ali, S.N. (1998) Molecular techniques in plant improvement programmes – A case for bambara groundnut (*Vigna subterranea* L. Verdc.). Paper presented at the Second International Workshop of BAMNET held at CSIR, Accra, Ghana, 23–25 September 1998.

Massawe, F.J., Azam-Ali, S.N. and Roberts, J.A. (2000) The use of molecular markers to explore phenotypic variation between and within landraces of bambara groundnut (*Vigna subterranea* L. Verdc.). Abstract. *Journal of Experimental Botany* 51, 71.

Matthews, R.B., Azam-Ali, S.N. and Peacock, J.M. (1990a) Response of four sorghum genotypes to mid-season drought. 2. Leaf characteristics and solar radiation interception. *Field Crops Research* 25, 297-308.

Matthews, R.B., Reddy, D.M., Rani, A.U. and Azam-Ali, S.N. (1990b) Response of four sorghum genotypes to mid-season drought. 1. Dry matter production, yield and water use. *Field Crops Research* 25, 279–296.

Mead, R., Curnow, R.N. and Hasted, A.M. (1993) *Statistical Methods in Agriculture and Experimental Biology*. Chapman and Hall, London.

Mian, M.A.R., Bailey, M.A., Ashley, D.A., Wells, R., Carter, T.E. Jr., Parrott, W.A. and Boerma, H.R. (1996) Molecular markers associated with water use efficiency and leaf ash in soybean. *Crop Science* 36, 1252–1257.

Millin, D.J. (1987) *Factors Affecting the Quality of Tea. Quality Control in the Food Industry*, Vol. 4. Academic Press, London, pp. 127–160.

Monsi, M. and Saeki, T. (1953) Uber den Lichtfaktor in den Pflanzengesellschaften und seine bedeutung fur die Stoffproduktion. *Japanese Journal of Botany* 14, 605–614.

Monteith, J.L. (1965) Light distribution and photosynthesis in field crops. *Annals of Botany* 29, 17-37.

Monteith, J.L. (1972) Solar radiation and productivity in tropical ecosystems. *Journal of Applied Ecology* 9, 747–766.

Monteith, J.L. (1977) Climate and the efficiency of crop production in Britain. *Philosophical Transactions of the Royal Society, London* B281, 277–294.

Monteith, J.L. (1978) Reassessment of maximum growth rates for C3 and C4 crops. *Experimental Agriculture* 14, 1–5.

Monteith, J.L. (1981) Climate variation and growth of crops. *Quarterly Journal of Royal Meteorological Society* 107, 749–774.

Monteith, J.L. (1986a) Significance of the coupling between saturation vapour pressure deficit and rainfall in monsoon climates. *Experimental Agriculture* 22, 329-338.

Monteith, J.L. (1986b) How do crops manipulate water supply and demand? *Philosophical Transactions of the Royal Society, London* A316, 245–259.

Monteith, J.L. (1990) Conservative behaviour in the response of crops to water and light. In: Rabbinge, R., Goudriaan, J., van Keulen, H., Penning de Vries, F.W.T. and Van Laar, H.H. (eds) *Theoretical Production Ecology: Reflections and Prospects*. Pudoc, Wageningen, pp. 3–14.

Monteith, J.L. and Unsworth, M.H. (1990) *Principles of Environmental Physics*. Edward Arnold, London.

Monteith, J.L., Marshall, B., Saffell, R.A., Clarke, D., Gallagher, J.N., Gregory, P.J., Ong, C.K., Squire, G.R. and Terry, A. (1983) Environmental control of a glasshouse suite for crop physiology. *Journal of Experimental Botany* 34, 309-321.

Morgan, J.M., Hare, R.A. and Fletcher, R.J. (1986) Genetic variation in osmoregulation in bread and durum wheats and its relationship to grain yields in a range of field environments. *Australian Journal of Agricultural Research* 37, 449–457.

Morgan, W.C. (1989) Alternatives to herbicides. *Plant Protection Quarterly* 4, 33-37.

Morris, R.A. and Garrity, D.P. (1993) Resource capture and utilisation in intercropping: non-nitrogen nutrients. *Field Crops Research* 34, 319-334.

Moss, S.R. and Rubin, B. (1993) Herbicide-resistant weeds: a worldwide perspective. *Journal of Agricultural Science* 120, 141–148.

Muchow, R.C. (1985) An analysis of the effects of water-deficits on grain legumes grown in a semi-arid tropical environment in terms of radiation interception and its efficiency of use. *Field Crops Research* 11, 309-323.

Muchow, R.C. (1988a) Effect of nitrogen supply on the comparative productivity of maize and sorghum in a semi-arid tropical environment. 1. Leaf growth and leaf nitrogen. *Field Crops Research* 18, 1–16.

Muchow, R.C. (1988b) Effect of nitrogen supply on the comparative productivity of maize and sorghum in a semi-arid tropical environment. 3. Grain yield and nitrogen accumulation. *Field Crops Research* 18, 31–43.

Muchow, R.C. (1992) Effect of water and nitrogen supply on radiation interception and biomass accumulation of kenaf (*Hibiscus cannabiscus*) in a semi-arid tropical environment. *Field Crops Research* 28, 281–293.

Muchow, R.C. and Davis, R. (1988) Effect of nitrogen supply on comparative productivity of maize and sorghum in a semi-arid tropical environment. II. Radiation interception and biomass accumulation. *Field Crops Research* 18, 17–30.

Muchow, R.C., Coates, D.B., Wilson, G.L. and Foale, M.A. (1982) Growth and productivity of irrigated *Sorghum bicolor* (L.Moench) in Northern Australia. I. Plant density and arrangement effects on light interception and distribution and grain yield in the hybrid Texas 610R in low and medium latitudes. *Australian Journal of Agricultural Research* 33, 773–784.

Musembi, D.K. and Griffiths, J.F. (1986) The use of precipitation data to identify soil-moisture patterns and the growing seasons in Eastern Kenya. *Agricultural and Forest Meteorology* 37, 47–61.

Myers, R.J.K. (1980) The root system of a grain sorghum crop. *Field Crops Research* 3, 53–64.

Nader, H.M. and Faught, W.A. (1984) Maize yield response to different levels of nitrogen and phosphorus fertiliser application: a seven-season study. *East African Agriculture and Forestry Journal* 44, 147–156.

Nageswara Rao, R.C., Simmonds, L.P., Azam-Ali, S.N. and Williams, J.H. (1989a) Population, growth and water use of groundnut maintained on stored water. I. Growth of roots and shoots. *Experimental Agriculture* 25, 51–61.

Nageswara Rao, R.C., Williams, J.H. and Singh, M. (1989b) Genotypic sensitivity to drought and yield potential of peanut. *Agronomy Journal* 81, 887–893.

Natarajan, M. and Willey, R.W. (1980a) Sorghum–pigeon pea intercropping and the effects of plant population density. 1. Growth and yield. *Journal of Agricultural Science* 95, 51–58.

Natarajan, M. and Willey, R.W. (1980b) Sorghum–pigeon pea intercropping and the effects of plant population density. 2. Resource use. *Journal of Agricultural Science* 95, 59–65.

Nelder, J.A. (1962) New kinds of systematic designs for spacing experiments. *Biometrics* 18, 283-307.

Newbury, J. and Ford-Lloyd, B. (1999) The production of user-friendly molecular marker for studying plants. *Biotechnology News* 40, 5–6.

Nguyen, H.T., Babu, R.C. and Blum, A. (1997) Breeding for drought resistance in rice: Physiology and molecular genetic considerations. *Crop Science* 37, 1426–1434.

Nieuwolt, S. (1977) *Tropical Climatology: An Introduction to the Climates of the Low Latitudes.* John Wiley & Sons, London.

Ofori, F. and Stern, W.R. (1987) The combined effects of nitrogen fertiliser and density of the legume component on production efficiency in a maize/cowpea intercrop system. *Field Crops Research* 16, 43–52.

Ong, C.K. (1991) Interactions of light, water and nutrients in agroforestry systems. In: Avery, M.E., Cannell, M.G.R. and Ong, C.K. (eds) *Biophysiological Research for Asian Agroforestry*, Winrock International, New Delhi, pp. 107–124.

Ong, C.K. and Huxley, P. (eds) (1996) *Tree–Crop Interactions. A Physiological Approach.* CAB International, Wallingford, 385 pp.

Ong, C.K., Corlett, J.E., Singh, R.P. and Black, C.R. (1991) Above and below-ground interactions in agroforestry systems. *Forest Ecology and Management* 45, 45–57.

Ong, C.K., Black, C.R., Marshall, F.M. and Corlett, J.E. (1996) Principles of resource capture and utilization of light and water. In: Ong, C.K. and Huxley, P. (eds) *Tree–Crop Interactions. A Physiological Approach.* CAB International, Wallingford, pp. 73–158.

Orozco-Castillo, C., Chalmers, K.J., Waugh, R. and Powell, W. (1994) Detection of genetic diversity and selective gene introgression in coffee using RAPD markers. *Theoretical and Applied Genetics* 87, 934–940.

Oswald, A., Abayo, G., Ransom, J.K., Kroschel, J. and Sauerborn, J. (1997) Catch-cropping with Sudan grass – an option for Striga control in subsistence agriculture. *Proceedings of the Brighton Crop Protection Conference – Weeds.* British Crop Protection Council, Farnham, UK, pp. 227–232.

Overseas Development Adminisration (1987) *Microclimatology in Tropical Agriculture,*

Vol 1. Final Report, research schemes R3208 and R3819. ODA, London.

Paffen, K. (1967) Das Verhaltnis der tages- zur jahreszeitlichen Temperaturschwankung, *Erdkunde* 21, 94–111.

Passmore, R. and Eastwood, M.A. (eds) (1986) *Human Nutrition and Dietetics*. Churchill Livingstone, Edinburgh.

Peacock, J.M., Azam-Ali, S.N. and Matthews, R.B. (1988) An approach to screening for resistance to water and heat stress in sorghum (*Sorghum bicolor* (L.) Moench). In: Whitehead, E.E., Hutchinson, C.F., Timmerman, B.N. and Varady, R.G. (eds) *Arid Lands Today and Tomorrow*. Proceedings of the International Research and Development Conference, 20–25 October, 1985, Tucson, Arizona. Westview Press, Boulder, Colorado, pp. 487–497.

Peacock, J.M., Miller, W.B., Matsuda, K. and Robinson, D.L. (1990) Role of heat girdling in early seedling death of sorghum. *Crop Science* 30, 138–143.

Peacock, J.M., Soman, P., Jayachandran, R., Rani, A.U., Howarth, C.J. and Thomas, A. (1993) Effects of high soil surface temperature on seedling survival in pearl millet. *Experimental Agriculture* 29, 215–225.

Penman, H.L. (1948) Natural evaporation from open water, bare soil and grass. *Proceedings of the Royal Society of London, Series A*, 193, pp. 120–145.

Penning de Vries, F.W.T., Jansen, D.M., ten Berge, H.F.M. and Bakema, A. (1989) *Simulation of Ecophysiological Processes of Growth in Several Annual Crops*. Pudoc, Wageningen.

Pons, T.L., van Riijnberk, H., Scheurwater, I. and van der Werf, A. (1993) Importance of the gradient in photosynthetically active radiation in a vegetation stand for leaf nitrogen allocation in two monocotyledons. *Oecologia* 95, 416–424.

Poulter, N.H. and Caygill, J.C. (1980) Vegetable milk processing and rehydration characteristics of bambara groundnut (*Voandzeia subterranea* L. Thouars). *Journal of the Science of Food and Agriculture* 31, 1158–1163.

Raper, C.D. and Downs, R.J. (1976) Field phenotype in phytotron culture: a case study for tobacco. *The Botanical Review* 42, 317–343.

Reddy, M.S. and Willey, R.B. (1981) Growth and resource use studies in an intercrop of pearl millet/groundnut. *Field Crops Research* 4, 13–24.

Rees, A.R. (1963) Relationship between growth rate and leaf area index in the oil palm. *Nature* 197, 63–64.

Reeves, M. (1992) The role of VAM fungi in nitrogen dynamics in maize–bean intercrops. *Plant Soil* 144, 85–92.

Riggs, T.J., Hansen, P.R., Start, N.D., Miles, D.M., Morgan, C.L. and Ford, M.A. (1981) Comparsion of spring barley varieties grown in England and Wales between 1880 and 1980. *Journal of Agricultural Science* 97, 599–610.

Rosielle, A.A. and Hamblin, J. (1981) Theoretical aspects of selection for yield in stress and non-stress environments. *Crop Science* 21, 943–946.

Ruthenburg, H. (1980) *Farming Systems in the Tropics*, 3rd edn. Clarendon Press, Oxford, 424 pp.

Ryel, R.J., Barnes, P.W., Beyschlag, W., Caldwell, M.M. and Flint, S.D. (1990) Plant competition for light analyzed with multispecies canopy model. 1. Model development and influence of enhanced UV-B conditions on photosynthesis in mixed wheat and wild oat canopies. *Oecologia* 82, 304-310.

Sagar, G.R. and Mortimer, A.M. (1976) An approach to the study of the population dynamics of plants with special reference to weeds. *Applied Biology* 1, 1–46.

Sellers, W.D. (1965) *Physical Climatology*. University of Chicago Press, Chicago, Illinois.

Sellschop, J.P.F. (1962) Cowpeas, *Vigna unguiculata* (L) Walp. *Field Crop Abstracts* 15, 259–266.

Sesay, A., Saboleh, S. and Yarmah, A. (1997) Farmers knowledge and cultivation of bambara groundnut in Sierra Leone. In: *Proceedings of the International Bambara Groundnut Symposium*, University of Nottingham, UK.

Sesay, A., Kunene, I. and Earnshaw, D.M. (1999) Bambara groundnut (*Vigna subterranea*) cultivation in Swaziland. Report of a farmers survey, University of Swaziland, Kwaluseni.

Sharma, S.K., Dawson, I.K. and Waugh, R. (1995) Relationships among cultivated and wild lentils revealed by RAPD analysis. *Theoretical and Applied Genetics* 91, 647–654.

Simmonds, L.P. (1979) *Principles of Crop Improvement*. Longman, London.

Simmonds, L.P. and Azam-Ali, S.N. (1989) Population, growth and water use of groundnut maintained on stored water. IV. The influence of population on water supply and demand. *Experimental Agriculture* 25, 87–98.

Simmonds, N.W. (1981) Genotype (G), Environment (E) and GE components of crop yields. *Experimental Agriculture* 17, 355-362.

Simmonds, N.W. and Smartt, J. (1999) *Principles of Crop Improvement.* Blackwell Science, Oxford, 412pp.

Sinclair, T.R. and Ludlow, M.M. (1985) Who taught plants thermodynamics? The unfulfilled potential of water potential. *Australian Journal of Plant Physiology* 12, 213–218.

Sinclair, T.R. and Muchow, R.C. (1999) Radiation use efficiency. *Advances in Agronomy* 65, 215–265.

Singh, A.K., Smartt, J., Simpson, C.E. and Raina, S.N. (1998) Genetic variation vis-à-vis molecular polymorphism in groundnut, *Arachis hypogaea* L. *Genetic Resources and Crop Evolution* 45, 119–126.

Sinoquet, H. and Caldwell, R.M. (1995) Estimation of light capture and partitioning in intercropping systems. In: Sinoquet, H. and Cruz, P. (eds) *Ecophysiology of Tropical Intercropping.* INRA, Paris, pp. 79–97.

Sivakumar, M.V.K. and Virmani, S.M. (1984) Growth and resource use of maize, pigeonpea and maize/pigeonpea intercrop in an operational research watershed. *Experimental Agriculture* 16, 377-386.

Slatyer, R.O. (1967) *Plant Water Relationships.* Academic Press, London.

Smartt, J. and Simmonds, N.W. (1995) *Evolution of Crop Plants*, 2nd edn. John Wiley & Sons, Chichester, UK, 531pp.

Smith, M.C. and Webb, M. (1993) A survey of weed problems in a region of Mali. *Proceedings of the Brighton Crop Protection Conference – Weeds.* British Crop Protection Council, Farnham, pp. 155–1160.

Soussana, J.F. and Arregui, M.C. (1995) Effect of mixed cropping on the nitrogen nutrition and growth of perennial ryegrass and white clover. *Agronomie* 15, 81–96.

Spink, J.H., Whaley, J., Semere, T., Wade, A., Sparkes, D. and Foulkes, J. (2000) *Prediction of Optimum Plant Population in Winter Wheat.* Home Grown Cereals Project Report No. 234. HGCA, Caledonia House, London.

Squire, G.R. (1979) Weather, physiology and seasonality of tea (*Camellia sinensis*) yields in Malawi. *Experimental Agriculture* 15, 321–330.

Squire, G.R. (1986) A physiological analysis for oil palm trials. *PORIM Bulletin* 12, pp.12–31.

Squire, G.R. (1990) *The Physiology of Tropical Crop Production.* CAB International, Wallingford.

Squire, G.R. and Callander, B.A. (1981) Tea plantations. In: Kozlowski, T.T. (ed.) *Water Deficits and Plant Growth, VI.* Academic Press, New York, pp. 471–510.

Squire, G.R. and Corley, R.H.V. (1987) Oil palm. In: Sethuraj, M.R. and Raghavendra, A.S. (eds) *Tree Crop Physiology.* Elsevier, Amsterdam, pp. 141–167.

Squire, G.R. Obaga, S.M.O. and Othieno, C.O. (1993) Altitude, temperature and shoot production of tea in the Kenyan Highlands. *Experimental Agriculture* 29, 107–120.

Squire, G.R. and Gibson, G.J. (1997) Scaling up and down: matching research with requirements in land management and policy. In: van Gardingen, P. (ed.) *Scaling Up.* Cambridge University Press, Cambridge, pp. 17–34.

Squire, G.R., Connolly, H., Crawford, J., Collinson, S.T. and Sesay, A. (1997) Linking vegetative and reproductive trait variability in landraces of bambara groundnut (*Vigna subterranea* L. Verdc). *Proceedings of the International Symposium on Bambara groundnut*, 23–25 July 1996, University of Nottingham, UK.

Stanton, W.R., Doughty, J., Orraca-Tetteh, R. and Steele, W. (1966) *Grain Legumes in Africa.* FAO, Rome.

Steduto, P. (1996). Water use efficiency. In: Pereira, L.S., Feddes, R.A., Gilley, J.R. and Lesaffre, B. (eds) *Sustainability of Irrigated Agriculture*, NATO ASI Series E: Applied Sciences. Kluwer Academic Publishers, Dordrecht, pp. 193–209.

Steduto, P and Hsiao, T.C. (1994). Radiation use efficiency of maize under well watered and water deficit conditions: cumulative vs. finite increment approach. *Proceedings of the 3rd European Society of Agronomy Meeting*, Abano-Padova, Italy, 18–22 Sept. 1994. ESA, BP52, Colmar Cedex, France, pp. 418–419.

Stigter, C.J. and Baldy, Ch.M. (1995) Manipulation of the microclimate by intercropping: making the best use of services rendered. In: Sinoquet, H. and Cruz, P. (eds) *Ecophysiology of Tropical Intercropping.* INRA, Paris, pp. 29–44.

Stockle, O.S. and Kiniry, J.R. (1990) Variability in crop radiation-use efficiency associated with vapour-pressure deficit. *Field Crops Research* 25, 171–181.

Stoorvogel, J.J. and Smaling, E. (1992) *Assessment of Soil Nutrient Depletion in Sub-Saharan Africa.* Winand Staring Centre, Wageningen.

Stott, P. (1998) Biogeography and ecology in crisis: the urgent need for a new metalanguage. *Journal of Biogeography* 25, 1–2.

Summerfield, R.J., Roberts, E.H., Ellis, R.H. and Lawn, R.J. (1991) Towards the reliable prediction of time to flowering in six annual crops. I. The development of simple models for fluctuating field environments. *Experimental Agriculture* 27, 11-31.

Syers, J.K. (1998) Managing soils for long-term productivity. In: Greenland, D.J., Gregory, P.J. and Nye, P.H. (eds) *Land Resources:on the Edge of a Malthusian Precipice?* CAB International, Wallingford, pp. 151–161.

Szeicz, G. (1974) Solar radiation in canopies. *Journal of Applied Ecology* 11, 1117–1156.

Tanner, C.B. and Sinclair, T.R. (1983) Efficient water use in crop production. In: Taylor, H.M., Jordan, W.R. and Sinclair, T.R. (eds) *Limitations to Efficient Water Use.* American Society of Agronomy, Madison, Wisconsin, pp. 1–27.

Tanton, T.W. (1979) Some factors limiting yields of tea (*Camellia sinensis*). *Experimental Agriculture* 15, 187–191.

Tanton, T.W. (1982a) Environmental factors affecting the yield of tea (*Camellia sinensis*). I. Effects of air temperature. *Experimental Agriculture* 18, 47–52.

Tanton, T.W. (1982b) Environmental factors affecting the yield of tea (*Camellia sinensis*). II. Effects of soil temperature, daylength and dry air. *Experimental Agriculture* 18, 53–63.

Tao, Y., Manners, J.M., Ludlow, M.M. and Hanzel, R.G. (1993) DNA polymorphism in grain sorghum (*Sorghum bicolor* L. Moench). *Theoretical and Applied Genetics* 86, 679–688.

Thornthwaite, C.W. (1948) An approach towards a rational classification of climate. *Geographical Review* 38, No.1.

Tournebize, R. and Sinoquet, H. (1995) Light interception and partitioning in a shrub/grass mixture. *Agricultural and Forest Meteorology* 72, 277–294.

Trenbath, B.R. (1986) Resource use by intercrops. In: Francis, C.A. (ed.) *Multiple Cropping Systems.* MacMillan, New York, pp. 57–81.

Troll, C. (1964) Karte der Jahreszeitenklimate der Erde. *Erdkunde* 18, 5–28.

Turner, N.C. (1981) Techniques and experimental approaches for the measurement of plant water status. *Plant Soil* 58, 339-366.

United Nations (1985) *The Least Developed Countries.* United Nations Conference on Trade and Development, United Nations, New York.

Van Kessel, C.H. and Roskoski, J.P. (1988) Row spacing effects on N_2-fixation, N-yield and soil N uptake of intercropped cowpea and maize. *Plant Soil* 111, 17–23.

Van Kessel, C., Singleton, P.W. and Hoben, H.J. (1985) Enhanced n-transfer from a soybean to maize by vesicular arbuscular mycorrhizal (vam) fungi. *Plant Physiology* 79, 562–563.

Van Keulen, H. and Wolf, J. (1989) *Modelling of Agricultural Production: Weather, Soils and Crops.* Simulation Monographs, Centre for World Food Studies, Pudoc, Wageningen.

Van Noordwijk, M., Lawson, G., Soumare, A., Groot, J.J.R. and Hairiah, K. (1996) Root distribution of trees and crops: competition and/or complementarity. In: Ong, C.K. and Huxley, P. (eds) *Tree–Crop Interactions. A Physiological Approach.* CAB International Wallingford, pp. 319-364.

Van Noordwijk, M., Cerri, C., Woomer, P.L., Nugroho, K. and Bernoux, M. (1997) Soil carbon dynamics in the humid tropical forest zone. *Geoderma* 79, 187–225.

Vavilov, N.I. (1928) *Geographical Centres of our Cultivated Plants.* Proceedings of the Fifth International Congress on Genetics, New York, pp. 342–369.

Wacquant, J.P., Ouknider, M. and Jacquard, P. (1989) Evidence for a periodic excretion of nitrogen by roots of grass-legume associations. *Plant Soil* 116, 57–68.

Wade, L.J., Myers, R.J.K. and Foale, M.A. (1991) Optimising plant stand in response to climatic risk. In: Muchow, R.C. and Bellamy, J.A. (eds) *Climatic Risk in Crop Production: Models and Management for the Semiarid Tropics and Subtropics.* CAB International, Wallingford, pp. 263–282.

Wahau, T.A.T. and Miller, D.A. (1978) Relative yield totals and yield components of intercropped sorghum and soybeans. *Agronomy Journal* 70, 287–291.

Wallace, J.S., Batchelor, C.H., Dabeesing, D.N. and Soopramanien, G.C. (1990) The partitioning of light and water in drip irrigated plant cane with a maize intercrop. *Agricultural Water Management* 17, 235–256.

White, J.W., Ochoa, M., Ibarra, P. and Singh, S.P. (1994) Inheritance of seed yield, maturity and seed weight of common bean (*Phaseolus vulgaris*) under semi-arid rainfed conditions. *Journal of Agricultural Science* 122, 265–273.

Wild, A. (ed.) (1988) *Russell's Soil Conditions and Plant Growth,* 11th edn. Longman, London.

Willey, R.W. (1979) Intercropping – its importance and research needs. II. Agronomy and research approach. *Field Crop Abstracts* 32, 78–85.

Willey, R.W. (1985) Evaluation and presentation of intercropping advantages. *Experimental Agriculture* 21, 119–133.

Willey, R.W. and Heath, S.B. (1969) The quantitative relationship between plant population and crop yield. *Advances in Agronomy* 21, 281–321.

Willey, R.W. and Osiru, D.S.O. (1972) Studies of mixtures of maize and bean (*Phaseolus vulgaris*) with particular reference to plant population. *Journal of Agricultural Science* 79, 517–524.

Willey, R.W. and Rao, M.R. (1981) Genotype studies at ICRISAT. In: *Proceedings of the International Workshop on Intercropping, 10–13 January 1979, ICRISAT, Patancheru, Andhra Pradesh, India*, pp. 105–116.

Williams, J.B. (1979) *Physical Aspects of Water Use Under Traditional and Modern Irrigation/Farming Systems in the Wadi Rima' Tihamah.* Land Resources Development Centre, Overseas Development Natural Resources Institute, Central Avenue.

Wilson, E.O. (1995) The current state of biological diversity. In: Wilson, E.O. and Peter, F.M. (eds) *Biodiversity*, 12th edn. National Academy Press, Washington, DC.

Wong, S.C., Cowan, I.R. and Farquhar, G.D. (1979) Stomatal conductance correlates with photosynthetic capacity. *Nature* 282, 424–426.

Wood, D. (1998) Ecological principles in agricultural policy: but which principles? *Food Policy* 5, 371–381.

Woodward, F.I. and Sheehy, J. (1983) *Principles and Measurements in Environmental Biology.* Butterworths, London.

Wust, G. (1922) Verdunstung und Niederschlag auf der Erde, *Zeitschrift der Gesellschaft f. Erdkunde*, Berlin, pp. 35–43.

Zagorodsky, M. (1911) Die Erderbse (*Voandzeia subterranea* Thours) und ihre Verwertung als Futtermittel. *Der Tropenpflanzer* 15, 413–436.

Zeven, A.C. (1998) Landraces: a review of definitions and classifications. *Euphytica* 104, 127–139.

Zulu, E.D. (1989) Germination and early seedling growth of bambara groundnut in response to temperature, sowing depth and soil water. MSc thesis, University of Nottingham, UK.

Appendix I

Terminology and Units for Agronomy

This document has been compiled to give guidelines on the consistent definition and presentation of units, data and terminology in agronomy.

As it has proved difficult to decide what will not be of practical use, the contents have been made as complete as possible, including quantities mainly used in physics but which might be encountered when dealing with instrumentation. For ease of reference the quantities and units have been arranged in tables by general category, with some quantities appearing in more than one table. Where appropriate, each line of the tables contains recommended (usually SI) units, along with alternative units and the most useful conversion factors. Scientific notation (e.g. '1.23×10^3') is used for numbers, except where the meaning is clearer using ordinary numerals rather than exponents.

The key rule is that only units set out under the latest version of the International System of Units or Système International d'Unités (SI) should be used. Do not use imperial measures or the old CGS (centimetre–gramme–second) units. The seven SI base units are shown in Table 1. These base units, and the official units which are derived from them, are used in Table 3, which shows recommended units for use in agronomy. Note that the SI system is coher-

ent, so that no numerical multiplication factor is needed to convert from the base units to the SI derived units. Some units such as the minute (min), hour (h), day (d), degree (°), minute ('), second ("), litre (l) and tonne (t) are non-SI units which need a numerical multiplication factor to convert them into SI base units, but they are officially recognized for use with the SI system where required. The hectare (ha) is another non-SI unit with widespread acceptance. Further information can be found in Baron (1994) and Salisbury (1996).

The material contained in this section is not exhaustive nor are the definitions universally accepted. However, wherever possible, we have attempted to be consistent with the scientific consensus or identified where our definitions differ from the general literature.

The book by Salisbury (1996) is recommended as an invaluable reference for units and terminology in plant physiology and also has useful guidance on scientific writing. Baron (1994) is another useful source for symbols and abbreviations, as well as editorial conventions. Readers may also find the reviews by Monteith (1984) and Czarnowski (1990) to be useful. We acknowledge the contribution of Dr Douglas Grindlay in the preparation of this appendix.

Table 1. The seven SI base units. Note that all SI derived units (e.g. joule, pascal, volt) can be expressed in terms of these base units without the introduction of any numerical factors.

Physical quantity and symbol	Unit name	Unit symbol
Length (l)	Metre or meter	m
Mass (m)	Kilogram[a,b,c]	kg
Time (t)	Second	s
Electric current (I)	Ampere	A
Thermodynamic temperature (T)	Kelvin	K (*not* °K)
Luminous intensity (I)	Candela	cd
Amount of substance (n, Q)	Mole[d]	mol

[a] *Not* kilogramme, which is the old name under the CGS system.
[b] Although it incorporates a prefix, for historical reasons the kilogram is the SI base unit, not the gram. Nevertheless, prefixes for multiples and sub-multiples are still attached to the gram.
[c] The kilogram is a unit of *mass* rather than *weight*, but for most purposes weight can be used as an acceptable synonym for mass. The *mass* of an object is a measure of the amount of matter it contains. Technically speaking, *weight* is a measure of the *force* produced by gravity (measured in Newtons), which varies slightly from place to place according to the strength of the earth's gravity. It is actually weight that is measured by typical electronic "balances", but mass can be measured if the balance has been calibrated with an object of known mass in that particular location.
[d] One mole of a substance contains Avogadro's number of particles ($\approx 6.022045 \times 10^{23}$ particles).

Table 2. Preferred prefixes for multiples and sub-multiples of SI and SI derived units. Multiples or sub-multiples should preferably be derived from the unit in steps of 10^3, but prefixes shown shaded are often used when convenient (e.g. cm).

Factor	Prefix	Symbol
10^{12}	tera	T
10^9	giga	G
10^6	mega	M
10^3	kilo	k
10^2	hecto	h
10^1	deca	da
10^{-1}	deci	d
10^{-2}	centi	c
10^{-3}	milli	m
10^{-6}	micro	µ
10^{-9}	nano	n
10^{-12}	pico	p
10^{-15}	femto	f
10^{-18}	atto	a

Table 3. A. Spatial

Physical quantity and symbol	Unit name	Unit symbol	Unit derivation	Other units and conversions, notes
Length (l)	Metre (meter)	m		centimetre (cm) = 0.01 m ångström (Å) = 0.1 nm (nanometre) = 10^{-10} m micron (μ) = 1 μm (micrometre) = 10^{-6} m inch (in) = 0.02540 m = 2.540 cm = 25.40 mm foot (ft) = 12 in = 0.3048 m = 30.48 cm yard (yd) = 3 ft = 36 in = 0.9144 m mile (mile) = 1760 yd = 1609.344 m = 1.609 344 km
Area (A)	Square metre (*not* 'metre squared')	m^2	m m	square centimetre (cm^2) = 10^{-4} m^2 1 m^2 = 10,000 cm^2 square inch (in^2) = 6.4516×10^{-4} m^2 = 6.4516 cm^2 square foot (ft^2) = 0.092 9030 m^2 = 929.030 cm^2 square yard (yd^2) = 0.836 127 m^2 = 8361.27 cm^2
Area – land	Hectare (*Derived from the obsolete unit the* are) (1 are = 100 m^2)	ha	10^4 m^2	1 ha = 100 m × 100 m = 0.01 km^2 = 2.471 05 acre acre (acre) = 4840 yd^2 = 4046.86 m^2 = $4.046\ 86 \times 10^{-3}$ km^2 = 0.404 686 ha = 1.563×10^{-3} square mile
	square kilometre	km^2	10^6 m^2	1 km^2 = 1000 m × 1000 m = 100 ha = 247.105 acre square mile ($mile^2$) = 640 acre = 2.589 99 $\times 10^6$ m^2 = 2.589 99 km^2 = 258.999 ha
Volume (V)	cubic metre (*not* 'metre cubed')	m^3	m m m	cubic centimetre (cm^3; *avoid* 'cc') = 10^{-6} m^3 cubic inch (in^3) = $1.638\ 71 \times 10^{-5}$ m^3 cubic foot (ft^3) = 0.028 3168 m^3 cubic yard (yd^3) = 0.764 555 m^3

A. Spatial (continued)

Physical quantity and symbol	Unit name	Unit symbol	Unit derivation	Other units and conversions, notes
Volume – capacity	Litre *The litre is not an SI unit but is accepted for use with the SI system*	l (or *l* or L) *The symbol L is preferable as l can be confused with the numeral 1*	10^{-3} m^3	m^3 = 1000 l 1 litre = 1000 cm^3 = 1 dm^3 millilitre (ml) = 10^{-3} l = 10^{-6} m^3 = 1 cm^3 hectolitre (hl) = 100 l = 0.1 m^3 UK fluid ounce (fl oz) = 0.028 4130 l = 28.4130 ml = 2.841 30 $\times$ 10^{-5} m^3 UK pint (pt) = 0.568 261 l = 5.682 61 $\times$ 10^{-4} m^3 UK gallon (gal) = 4.546 09 l = 4.546 09 $\times$ 10^{-3} m^3 UK bushel (bu) = 8 gal = 36.3687 l = 0.036 3687 m^3 US gallon = 0.8327 gal = 3.7854 l = 3.7854 $\times$ 10^{-3} m^3 US bushel (bu) = 35.2391 l = 0.035 2391 m^3
Linear velocity; transport velocity; permeability	Metre per second	m s^{-1}		1 m s^{-1} = 3600 m h^{-1} = 8.6400 $\times$ 10^4 m day^{-1} cm s^{-1} = 0.01 m s^{-1} ft s^{-1} = 0.3048 m s^{-1} km h^{-1} = 0.277 778 m s^{-1} = 0.6214 mile h^{-1} mile h^{-1} (mph) = 0.447 040 m s^{-1} = 1.609 344 km h^{-1} 1 m day^{-1} = 1.157 407 $\times$ 10^{-5} m s^{-1} 1 m year^{-1} = 3.169 $\times$ 10^{-8} m s^{-1}
	Metre per day Metre per year	m day^{-1} m year^{-1}		
Linear acceleration (*a*)	Metre per second squared	m s^{-2}		ft s^{-2} = 0.3048 m s^{-2}
Rate of flow	Cubic metre per second	m^3 s^{-1}		cubic foot per second (ft^3 s^{-1}) = 2.831 $\times$ 10^{-2} m^3 s^{-1}
Plane angle	Radian	rad		degree (°) = (π/180) rad minute (') = (π/10,800) rad second (") = (π/648,000) rad
Angular velocity (ω)	Radian per second	rad s^{-1}		degree per second (° s^{-1}) = 1.745 33 $\times$ 10^{-2} rad s^{-1}
Angular acceleration (α)	Radian per second squared	rad s^{-2}		degree per second squared (° s^{-2}) = 1.745 33 $\times$ 10^{-2} rad s^{-2}
Solid angle	Steradian	sr		

B. Mass, substance, concentration

Physical quantity and symbol	Unit name	Unit symbol	Unit derivation	Other units and conversions, notes
Mass (m) 'Weight' is often incorrectly used in place of mass – see note [b] under Table 1	kilogram (not kilogramme, which is the old name under the CGS system)	kg		gram (g) = 0.001 kg ounce (oz) = 0.028 3495 kg = 28.3495 g pound (lb) = 16 oz = 0.453 592 37 kg = 453.592 37 g UK stone (stone) = 14 lb = 6.350 29 kg = 6350.29 g UK hundredweight (cwt) = 112 lb = 50.8023 kg
Mass (e.g. yield)	tonne	t	10^3 kg	1 t = 1 Mg (megagram) Mt = 10^6 t (million tonnes) UK ton (ton) = 20 cwt = 2240 lb = 1.016 05 t
Amount of substance (n)	mole	mol		If using mol it is necessary to define the type of particle
Density (p)	kilogram per cubic metre (not 'kilogram per metre cubed')	kg m^{-3}		g cm^{-3} = 1000 kg m^{-3} pound per cubic foot (lb ft^{-3}) = 0.160 185 kg m^{-3}
Concentration (c) – solids in liquids	kilogram per cubic metre	kg m^{-3}		g m^{-3} = 0.001 kg m^{-3} kg l^{-1} = 1000 kg m^{-3} g l^{-1} = 0.001 kg l^{-1} = 1 kg m^{-3}
	moles per cubic metre moles per litre	mol m^{-3} mol l^{-1}	1000 mol m^{-3}	1 mol m^{-3} = 1 mmol l^{-1} = 0.001 mol l^{-1} = 0.001 M Avoid use of molarity/molar solutions in terms of M. molar solution (M) = 1 mole per litre (mol l^{-1}) ounce per gallon (oz/gal) = 6.236 g l^{-1} ppm = 1 mg l^{-1} = 0.001 g l^{-1} = 10^{-6} kg l^{-1} 1% = 10 g l^{-1} = 10,000 ppm

C. Time

Physical quantity and symbol	Unit name	Unit symbol	Unit derivation	Other units and conversions, notes
Time (t)	Second	s		Minute (min) = 60 s Hour (h) = 3600 s Day = 1440 min = 86 400 s Julian year (yr) = 365.25 day = 8766 h = 5.2596 × 10⁵ min = 3.15576 × 10⁷ s
Thermal time/ accumulated temperature (Θ)	Degree–days	°C^d		*Base temperature (T_b) and how thermal time was calculated should always be given*
Periodic frequency (f, v)	Hertz	Hz	s^{-1}	
Rotational frequency	Per second	s^{-1}		Revolutions per second (rev s^{-1}) Revolutions per minute (rev min^{-1}, rpm)

D. Force, energy, pressure

Physical quantity and symbol	Unit name	Unit symbol	Unit derivation	Other units and conversions, notes
Force (F)	Newton	N	$m\ kg\ s^{-2}$	dyne (dyn) = 10^{-5} N kilogram–force* (kgf) = 9.807 N pound–force* (lbf) = 4.448 22 N *Derived from the gravitational force on (i.e. weight of) an object of the corresponding mass, using a standardized value for the acceleration due to gravity (g) in the equation F = m × g
Energy (E); work (W)	Joule	J	$N\ m,\ m^2\ kg\ s^{-2}$	erg (erg) = dyn cm = 10^{-7} J kilowatt–hour (kW h) = 3.6 MJ = 3.6×10^6 J calorie (cal) = 4.1855 J (at 15° C) British thermal unit (Btu) = $1.055\ 06 \times 10^3$ J
Power (P)/ rate of work	Watt	W	$J\ s^{-1},\ m^2\ kg\ s^{-3}$	watt = joule per second erg s^{-1} = 10^{-7} W horsepower (hp) = 745.700 W
Pressure (p)	Pascal	Pa	$N\ m^{-2},\ J\ m^{-3}$	dyn cm^{-2} = 0.1 Pa torr = mm Hg = 133.322 Pa bar (bar) = 10^5 Pa = 0.1 MPa (megapascal) millibar (mbar) = 100 Pa standard atmosphere (atm) = 760 torr = 760 mm Hg = $1.013\ 25 \times 10^5$ Pa = 0.101 325 MPa psi (lbf in^{-2}) = 6.89476×10^3 Pa
Partial pressure of gaseous species j (P_j)	Pascal	Pa	$N\ m^{-2},\ J\ m^{-3}$	See pressure above
Stress (σ)	Pascal	Pa	$N\ m^{-2},\ J\ m^{-3}$	dyn cm^{-2} = 0.1 Pa
Torque (T)	Newton metre	N m		
Surface tension (σ, T)	Newton per metre	N m^{-1}		dyn cm^{-1} = 0.001 N m^{-1}
Dynamic viscosity (η)	Pascal second	Pa s	$N\ s\ m^{-2}$	centipoise = 10^{-3} Pa s

E. Electricity, magnetism

Physical quantity and symbol	Unit name	Unit symbol	Unit derivation	Other units and conversions, notes
Current (I)	Ampere	A		
Potential; potential difference (V); EMF (E)	Volt	V	$W\,A^{-1}$, $J\,C^{-1}$	
Resistance (R)	Ohm	Ω	$V\,A^{-1}$	
Power (P)	Watt	W	$V\,A$, $J\,s^{-1}$	
Charge; quantity (Q)	Coulomb	C	$s\,A$	
Capacitance (C)	Farad	F	$C\,V^{-1}$	
Inductance	Henry	H	$Wb\,A^{-1}$	
Conductance (G)	Siemens	S	$A\,V^{-1}$, Ω^{-1}	
Resistivity	Ohm metre	$\Omega\,m$		
Frequency (f, v)	Hertz	Hz	s^{-1}	
Electric field strength	Volt per metre	$V\,m^{-1}$		
Magnetic flux (Φ)	Weber	Wb	$V\,s$	
Magnetic flux density (B)	Tesla	T	$Wb\,m^{-2}$	gauss (G) = 10^{-4} T

F. Light

Physical quantity and symbol	Unit name	Unit symbol	Unit derivation	Other units and conversions, notes
Luminous intensity (I)	Candela	cd		*As the candela and its derivatives the lumen and lux are based on the sensitivity of the human eye, they should not be used in plant science*
Luminous flux (Φ)	Lumen	lm	cd sr	phot (lm cm^{-2}) = 10^4 lx
Illuminance	Lux	lx	lm m^{-2}	footcandle (fc) = 10.7639 lx
Number of photons or quanta	Mole	mol		einstein (E) = 1 mole of photons
Radiant energy (Q, Q_e)	Joule	J		
Irradiation/ radiant energy fluence	Joule per square metre	J m^{-2}		MJ ha^{-1} = 10^2 J m^{-2}
Radiant energy flux/ radiant power (ϕ_e, W)	Watt / Joule per second	W / J s^{-1}	J s^{-1}	*watt = joule per second*
Radiant energy flux density/ irradiance (E_e)	Watt per square metre / Kilowatt per square metre / Joule per square metre second	W m^{-2} / kW m^{-2} / J m^{-2} s^{-1}	J m^{-2} s^{-1} / 1000 W m^{-2}	erg cm^{-2} s^{-1} = 1000 W m^{-2} = 1 kW m^{-2} / *watt per square metre = joule per square metre second*
Photon or quantum flux density/ quantum irradiance (E_p)	Micromoles per square metre second / Millimoles per square metre second	µmol m^{-2} s^{-1} / mmol m^{-2} s^{-1}		1 µmol m^{-2} s^{-1} = 0.001 mmol m^{-2} s^{-1} = 10^{-6} mol m^{-2} s^{-1} / 1 mmol m^{-2} s^{-1} = 1000 µmol m^{-2} s^{-1} = 10^{-3} mol m^{-2} s^{-1}
Radiant intensity	Watt per steradian	W sr^{-1}	J sr^{-1} s^{-1}	*This refers to the output of a light source – do not use 'intensity' when 'irradiance' is meant*
Frequency (f, υ)	Hertz	Hz	s^{-1}	
Wavelength (λ)	Metre / Nanometre	m / nm	/ 10^{-9} m	*For visible light*

G. Heat, Temperature

Physical quantity and symbol	Unit name	Unit symbol	Unit derivation	Other units and conversions, notes
Thermodynamic temperature (T)	Kelvin	K (*not* °K)		
Celsius temperature (t)	Degree Celsius (not *degree centigrade*)	°C	K − 273.15	*Use for environmental temperatures* Fahrenheit (°F) °C = (5/9 × °F − 32)
Temperature difference or interval (θ)	kelvin	K		
Thermal time/ accumulated temperature (Θ)	degree–days	°C^{d}		*Base temperature (T_b) and how thermal time was calculated should always be given*
Quantity of heat (Q)	Joule	J		
Thermal conductivity (k)	Watt per metre kelvin	W m^{-1} K^{-1}		
Specific heat capacity	Joule per kilogram kelvin	J kg^{-1} K^{-1}		J g^{-1} K^{-1} = 10^{-3} J kg^{-1} K^{-1}
Latent heat		J kg^{-1}		J g^{-1} = 10^{-3} J kg^{-1}
Entropy		J K^{-1}		

H. Atomic

Physical quantity and symbol	Unit name	Unit symbol	Unit derivation	Other units and conversions, notes
Radionuclide activity (A)	Becquerel	Bq	s^{-1}	Curie (Ci) = 3.7 × 10^{10} Bq
Absorbed dose	Gray	Gy	J kg^{-1}	rad (rad) = 0.01 Gy
Dose equivalent	Sievert	Sv	J kg^{-1}	rem (rem) = 0.01 Sv
Energy	Electronvolt	eV	1.6022 × 10^{-19} J	
Molecular mass/weight	Gram per mole	g mol^{-1}		

I. Photosynthesis, respiration

Physical quantity and symbol	Unit name	Unit symbol	Unit derivation	Other units and conversions, notes
Number of photons or quanta of *PAR*	Mole (*PAR*)	mol		einstein (E) = 1 mole of photons; *PAR* = 400–700 nm (0.4–0.7µm)
Photosynthetic energy	Joule (*PAR*)	J		
Photosynthetic energy fluence	Joule (*PAR*) per square metre	$J\ m^{-2}$		$MJ\ ha^{-1} = 10^2\ J\ m^{-2}$
Photosynthetic energy flux/Power	Watt (*PAR*); Joule (*PAR*) per second	W; $J\ s^{-1}$	$J\ s^{-1}$	*watt = joule per second*
Photosynthetic energy flux density/ Photosynthetic irradiance (*PI*)	Watt (*PAR*) per square metre; Joule (*PAR*) per square metre second	$W\ m^{-2}$; $J\ m^{-2}\ s^{-1}$	$J\ m^{-2}\ s^{-1}$	$erg\ cm^{-2}\ s^{-1} = 0.001\ W\ m^{-2}$; $1\ W\ m^{-2}\ PAR \approx 4.6\ \mu mol\ m^{-2}\ s^{-1}\ PAR$; $1\ J\ m^{-2}\ s^{-1}\ PAR \approx 4.6\ \mu mol\ m^{-2}\ s^{-1}\ PAR$
Photosynthetic photon flux density (*PPFD*)/ photosynthetic quantum irradiance (*PQI*)	Micromoles (*PAR*) per square metre second	$\mu mol\ m^{-2}\ s^{-1}$		$1\ \mu mol\ m^{-2}\ s^{-1} = 0.001\ mmol\ m^{-2}\ s^{-1}$ $= 10^{-6}\ mol\ m^{-2}\ s^{-1}$; $1\ \mu mol\ m^{-2}\ s^{-1}\ PAR \approx 0.218\ W\ m^{-2}\ PAR$
	Millimoles (*PAR*) per square metre second	$mmol\ m^{-2}\ s^{-1}$		$1\ mmol\ m^{-2}\ s^{-1} = 1000\ \mu mol\ m^{-2}\ s^{-1}$ $= 10^{-3}\ mol\ m^{-2}\ s^{-1}$; $1\ mmol\ m^{-2}\ s^{-1}\ PAR \approx 218\ W\ m^{-2}\ PAR$
Daily photosynthetic photon flux density	Moles (*PAR*) per square metre day	$mol\ m^{-2}\ day^{-1}$		
Fractional interception by canopy (*f*)	Proportion of incident radiation intercepted	unitless – decimal fraction *or* %	$J\ J^{-1}$ $mol\ mol^{-1}$ (× 100%)	
CO_2 concentration	Micromoles (CO_2) per mole (air)	$\mu mol\ mol^{-1}$		
CO_2 exchange rate (*CER*) – leaf	Micromoles (CO_2) per square metre (leaf area) second	$\mu mol\ m^{-2}\ s^{-1}$		Typical values 1–40 $\mu mol\ m^{-2}\ s^{-1}$

I. Photosynthesis, respiration (continued)

Physical quantity and symbol	Unit name	Unit symbol	Unit derivation	Other units and conversions, notes
Daily net CO_2 exchange rate – plants	Millimoles (CO_2) per square metre day	$mmol\ m^{-2}\ day^{-1}$		
Net photosynthetic rate (P_N)	Micromoles (CO_2) per square metre second	$\mu mol\ m^{-2}\ s^{-1}$		$mg\ dm^{-2}\ h^{-1}\ (CO_2) = 0.63\ \mu mol\ m^{-2}\ s^{-1}$
	Micromoles (CO_2) per kilogram (dry matter) second	$\mu mol\ kg^{-1}\ s^{-1}$		$mg\ g^{-1}\ h^{-1}\ (CO_2) = 6.3\ \mu mol\ kg^{-1}\ s^{-1}$
Quantum yield of photosynthesis	Moles (CO_2) per mole (PAR)	$mol\ mol^{-1}$		
Water use efficiency – CO_2 ($WUE = P_N/E$)	Millimoles (CO_2) per mole (water)	$mmol\ mol^{-1}$		
	Gram (CO_2) per kilogram (water)	$g\ kg^{-1}$		
Respiration rate (R)	Micromoles (CO_2) per square metre second	$\mu mol\ m^{-2}\ s^{-1}$		$mg\ dm^{-2}\ h^{-1}\ (CO_2) = 0.63\ \mu mol\ m^{-2}\ s^{-1}$
	Micromoles (CO_2) per kilogram (dry mass) second	$\mu mol\ kg^{-1}\ s^{-1}$		$mg\ g^{-1}\ h^{-1}\ (CO_2) = 6.3\ \mu mol\ kg^{-1}\ s^{-1}$

J. Plant growth analysis

Physical quantity and symbol	Unit name	Unit symbol	Unit derivation	Other units and conversions, notes
Dry mass or 'weight' (W); fresh mass or 'weight'	Kilogram per square metre	kg m^{-2}		
	Gram per square metre	g m^{-2}	10^{-3} kg m^{-2}	*Often more useful practically and relates to size of quadrats* 1 g m^{-2} = 10 kg ha^{-1} = 0.01 t ha^{-1} 1 t ha^{-1} = 1000 kg ha^{-1} = 100 g m^{-2}
	Tonne per hectare *If more than 999 kg ha^{-1}*	t ha^{-1}	10^{-1} kg m^{-2}	
Percent dry matter/mass/'weight'	(Dry mass/fresh mass) expressed as percentage	%	kg kg^{-1} × 100%	
Growth rate – dry mass or 'weight' (l)	Kilogram (dry mass) per day	kg day^{-1}		*Not SI base units, but often more useful practically*
	Gram (dry mass) per day	g day^{-1}		
Growth rate – linear (e.g. root length, root depth)	Centimetre per day	cm day^{-1}		*Not SI units, but often more useful practically*
	Millimetre per day	mm day^{-1}		
Thermal rate of linear extension (e.g. leaves)	Millimetre per degree-day	mm(°C^{d-1})	10^{-3} m (°C^{d-1})	
Root to shoot ratio	Kilogram (root) per kilogram (shoot)	kg kg^{-1} (i.e. unitless)		
Crop growth rate ($CGR = NAR \times LAI$) *Also given symbol Γ*	Kilogram per square metre (ground) day	kg m^{-2} day^{-1}		t ha^{-1} day^{-1} = 0.1 kg m^{-2} day^{-1} *Avoid use of weeks, as non–SI unit and of no physiological significance for plants*
	Gram per square metre (ground) day	g m^{-2} day^{-1}		*Not SI base units, but often more useful practically* *Typical values 1–40 g m^{-2} day^{-1}*
Relative growth rate ($RGR = NAR \times LAR$)	Kilogram (dry mass) per kilogram (dry mass) day	kg kg^{-1} day^{-1}		*Avoid g g^{-1} day^{-1}, as non-SI units*
	Gram (dry mass) per kilogram (dry mass) day	g kg^{-1} day^{-1}		1 g kg^{-1} day^{-1} = 0.001 kg kg^{-1} day^{-1} *Typical values 10–400 g kg^{-1} day^{-1}*

J. Plant growth analysis (continued)

Physical quantity and symbol	Unit name	Unit symbol	Unit derivation	Other units and conversions, notes
Net assimilation rate (*NAR*)	Kilogram (dry mass) per square metre (leaf area) day	kg m^{-2} day^{-1}		
	Gram (dry mass) per square metre (leaf area) day	g m^{-2} day^{-1}		*Not SI base units, but often more useful practically* Typical values 1–30 g m^{-2} day^{-1}
Plant leaf area	Square metre (leaf area) per plant	m^2		
Leaf area index (*LAI, L*)	Square metre (leaf area) per square metre (ground)	m^2 m^{-2} (i.e. unitless)		Typical values 0.01–10
Green area index (*GAI*)	Square metre (green area) per square metre (ground)	m^2 m^{-2} (i.e. unitless)		
Leaf area ratio (*LAR*)	Square metre (leaf area) per kilogram (pla<u>n</u>t dry mass)	m^2 kg^{-1}		Typical values 10–60 m^2 kg^{-1}
Specific leaf area (*SLA*)	Square metre (leaf area) per kilogram (<u>leaf</u> dry mass)	m^2 kg^{-1}		cm^2 g^{-1} = 0.1 m^2 kg^{-1} Typical values 5–30 m^2 kg^{-1}
Specific leaf mass/ Specific leaf weight (*SLW*)	Kilogram (leaf dry mass) per square metre (leaf area)	kg m^{-2}		Typical values 0.033–0.200 kg m^{-2}
Leaf mass ratio/ Leaf weight ratio	Kilogram (leaf dry mass) per kilogram (plant dry mass)	kg kg^{-1}		Typical values 0.4–0.8 kg kg^{-1}

K. Agronomy – see also Sections I, J and L

Physical quantity and symbol	Unit name	Unit symbol	Unit derivation	Other units and conversions, notes
Seed or grain moisture content (MC)	Percentage of grain mass	%	$kg\ kg^{-1} \times 100\%$	
Seed 'weight', grain 'weight'	Milligram per seed	mg	$10^{-6}\ kg$	
Thousand seed weight/ Thousand grain weight	Gram per thousand seeds	g	$10^{-3}\ kg$	
Grain specific weight/ Hectolitre weight	Kilogram per hectolitre	$kg\ hl^{-1}$	$10\ kg\ m^{-3}$	
Seed or grain protein	Percentage of grain mass	%	$kg\ kg^{-1} \times 100\%$	*Mass of grain protein : mass of grain N = 5.83:1*
Hagberg falling number (grain α-amylase)	Second	s		*Bread making > 250 s* / *Biscuit/cake making > 180 s*
Seed rate	Seeds per square metre	m^{-2}		seed per hectare $(ha^{-1}) = 10^{-4}\ m^{-2}$; $1\ m^{-2} = 10,000\ ha^{-1}$; seed per acre $(acre^{-1}) = 2.4711 \times 10^{-4}\ m^{-2}$ $= 2.4711\ ha^{-1}$
	Kilogram per hectare	$kg\ ha^{-1}$	$10^{-4}\ kg\ m^{-2}$	$g\ m^{-2} = 10\ kg\ ha^{-1}$
Plant population density (Np) (*not 'plant population'*) *Similar for ears, grains, tillers and shoots per square metre*	Plant per square metre	m^{-2}		plant per hectare $(ha^{-1}) = 10^{-4}\ m^{-2}$; $1\ m^{-2} = 10,000\ ha^{-1}$; plant per acre $(acre^{-1}) = 2.4711 \times 10^{-4}\ m^{-2}$ $= 2.4711\ ha^{-1}$
Grain yield	Tonne per hectare	$t\ ha^{-1}$	$10^{-1}\ kg\ m^{-2}$	$kg\ ha^{-1} = 0.001\ t\ ha^{-1}$; $g\ m^{-2} = 0.01\ t\ ha^{-1}$; pound per acre $(lb\ acre^{-1}) = 1.12085 \times 10^{-3}\ t\ ha^{-1}$; hundredweight per acre $(cwt\ acre^{-1}) = 0.1255\ t\ ha^{-1}$; ton per acre $(ton\ acre^{-1}) = 2.511\ t\ ha^{-1} = 2511\ kg\ ha^{-1}$

K. Agronomy (continued)

Physical quantity and symbol	Unit name	Unit symbol	Unit derivation	Other units and conversions, notes
Components of grain yield *Multiply together to grain yield in kg m^{-2} or kg ha^{-1}*	Ears per m^2 or per hectare m^{-2}, ha^{-1} Grains per ear mass or 'weight' per grain	mg kg		*Convert from mg:* 1 mg = 10^{-6} kg
Harvest index (*H*)	Unit mass (harvested, usable or edible yield) per unit mass (total yield)	unitless – decimal fraction or %	kg kg^{-1} (× 100%)	*Specify whether harvested, usable or edible yield is used*
Spray rate *See also concentration*	Litre per hectare Gram per hectare	l ha^{-1} g ha^{-1}	10^{-3} m^3 ha^{-1} 10^{-3} kg ha^{-1}	pint per acre (pt acre^{-1}) = 1.404 l ha^{-1} gallon per acre (gal acre^{-1}) = 11.24 l ha^{-1} ounce per acre (oz acre^{-1}) = 70.053 g ha^{-1}
Fertilizer rate *For N fertilizer rates, write as kg ha^{-1} N or kg ha^{-1} nitrogen, not kgN ha^{-1} or kgN/ha*	Kilogram per hectare	kg ha^{-1}	10^{-4} kg m^{-2}	g m^{-2} = 10 kg ha^{-1} t ha^{-1} = 1000 kg ha^{-1} pound per acre (lb acre^{-1}) = 1.121 kg ha^{-1} hundredweight per acre (cwt acre^{-1}) = 125.5 kg ha^{-1} ton per acre (ton acre^{-1}) = 2511 kg ha^{-1} = 2.511 t ha^{-1} unit per acre = 0.01 cwt acre^{-1} = 1.255 kg ha^{-1}
Nitrogen uptake (= dry mass (%N))	Kilogram per hectare Gram per square metre	kg ha^{-1} (kg ha^{-1} N) g m^{-2} (g m^{-2} N)	10^{-4} kg m^{-2} 10^{-3} kg m^{-2}	1 kg ha^{-1} = 0.1 g m^{-2} 1 g m^{-2} = 10 kg ha^{-1}
Nutrient concentration in plant tissue (e.g. N, P, K)	Mole per kilogram Gram per kilogram Percentage of dry mass	mol kg^{-1} g kg^{-1} %		*Specify fresh or dry mass Use of mol kg^{-1} is preferable*
Stem soluble carbohydrate	Kilogram per hectare Gram per square metre Tonne per hectare	kg ha^{-1} g m^{-2} t ha^{-1}	10^{-4} kg m^{-2} 10^{-3} kg m^{-2} 10^{-1} kg m^{-2}	

K. Agronomy (continued)

Physical quantity and symbol	Unit name	Unit symbol	Unit derivation	Other units and conversions, notes
Radiation use efficiency (RUE; ε_s)	Gram (dry mass) per megajoule intercepted	g MJ^{-1}		Also called dry matter/intercepted radiation ratio or quotient. Radiation use 'efficiency' is best avoided, as strictly speaking efficiency is a dimensionless value between 0 and 1 (same units on top and bottom of equation)
Daily solar energy absorbed by crop	Megajoule per square metre (ground) day	MJ m^{-2} day^{-1}		
Annual solar energy absorbed by crop	Gigajoule per square metre (ground) year	GJ m^{-2} yr^{-1}		
Fractional interception by canopy (f)	Proportion of incident radiation intercepted	unitless – decimal fraction or %	J J^{-1} or mol mol^{-1} ($\times$ 100%)	
Water use efficiency (WUE; ε_w)	Gram (dry mass) per kilogram (water transpired)	g kg^{-1}		mg g^{-1} = g kg^{-1}. Also called dry matter/transpired water ratio or quotient. Water use 'efficiency' is best avoided, as strictly speaking efficiency is a dimensionless value between 0 and 1 (same units on top and bottom of equation)
	Gram (dry mass) per millimetre (water transpired)	g mm^{-1}		1 g mm^{-1} = 1 g kg^{-1}
Water inflow rate per unit root length (I)	Gram (water) per metre of root per day	g m^{-1} day^{-1}		
Root length density (l_v)	Metre (root) per cubic metre (soil)	m m^{-3}		cm cm^{-3} = 10^4 m m^{-3}
Root length per unit ground area (R)	Kilometre (root) per square metre (ground)	km m^{-2}		

K. Agronomy (continued)

Physical quantity and symbol	Unit name	Unit symbol	Unit derivation	Other units and conversions, notes
Transpiration rate (E)	Micromoles (water) per square metre second	μmol m^{-2} s^{-1}		
Evapotranspiration rate (E_t)	Micromoles (water) per square metre second	μmol m^{-2} s^{-1}		
	Millimetre per day	mm day^{-1}	10^{-3} m day^{-1}	1 mm day^{-1} = 1 dm^3 m^{-2} day^{-1} = 1 l m^{-2} day^{-1} 1 mm day^{-1} $\approx$ 1 kg m^{-2} day^{-1} Inch per day (in day^{-1}) = 25.4 mm day^{-1}
Water potential (ψ), leaf water potential (ψ_l), solute potential (ψ_s), turgor potential (ψ_p), matric potential (ψ_m), gravitational potential (ψ_g)	Pascal Megapascal	Pa MPa		*Water potential usually negative* 1 MPa = 106 Pa
Metabolic energy (*ME*) content per unit dry mass (*D-value*)	Megajoule per kilogram (dry mass)	MJ kg^{-1}		kJ g^{-1} = MJ kg^{-1} *Typical D–value = 6.3 MJ kg^{-1} for most feeds other than oil seeds*
Metabolic energy (*ME*) content per unit ground area	Kilojoule per square metre Megajoule per hectare	kJ m^{-2} MJ ha^{-1}		1 kJ m^{-2} = 10 MJ ha^{-1} 1 MJ ha^{-1} = 0.1 kJ m^{-2}
Energy consumption – livestock	Megajoule per day	MJ day^{-1}		
Stocking density	Animals per hectare	ha^{-1}	10^{-4} m^{-2}	Animal per square metre (m^{-2}) = 10 000 ha^{-1}

L. Crop environment – see also Sections F, G and K

Physical quantity and symbol	Unit name	Unit symbol	Unit derivation	Other units and conversions, notes
Total/global irradiation	Megajoule per square metre (ground)	$MJ\ m^{-2}$		
Total/global irradiance (S)	Megajoule per square metre (ground) day	$MJ\ m^{-2}\ day^{-1}$		
Photoperiod (P)	Hour	h		
Thermal time/Accumulated temperature (Θ)	Degree–days	$^\circ C^d$		*Base temperature (T_b) and how thermal time was calculated should always be given*
Precipitation/rainfall, irrigation, amount of evaporation	Millimetre	mm	10^{-3} m	$1mm = 1\ dm^3\ m^{-2} = 1\ l\ m^{-2} = 10\ m^3\ ha^{-1}$ ($\approx 1\ kg\ m^{-2}$ or $10\ t\ ha^{-1}$ in terms of mass of water) inch (in) = 25.4 mm
Evaporation rate	Millimetre per day	$mm\ day^{-1}$	10^{-3} m day^{-1}	$1mm\ day^{-1} = 1\ dm^3\ m^{-2}\ day^{-1} = 1\ l\ m^{-2}\ day^{-1}$ $1mm\ day^{-1} \approx 1\ kg\ m^{-2}\ day^{-1}$ inch per day (in day^{-1}) = 25.4 mm day^{-1}
Water vapour pressure	Kilopascal	kPa		$1\ kPa = 10^3\ Pa$
Saturation water vapour pressure deficit/saturation deficit (D)	Kilopascal	kPa		$1\ kPa = 10^3\ Pa$
Relative humidity (RH)	Percent	% RH		
Wind speed (V)	Metre per second	$m\ s^{-1}$		mile h^{-1} (mph) = 0.447 040 m s^{-1}
Wind direction	Degrees	°		
Acidity/alkalinity	pH units	pH (plus value)		
Soil volumetric water content	Cubic metre (water) per cubic metre (soil)	$m^3\ m^{-3}$		$1\ m^3\ m^{-3} = 1\ m\ m^{-1}$
	Metre (water) per metre (soil)	$m\ m^{-1}$		$mm\ m^{-1} = 0.001\ m\ m^{-1}$

L. Crop environment (continued)

Physical quantity and symbol	Unit name	Unit symbol	Unit derivation	Other units and conversions, notes
Soil water potential *See water potential in Section K*	Kilopascal	kPa	10^3 Pa	*Negative*
Soil dry bulk density	Gram per cubic centimetre (dry soil)	g cm^{-3}	10^3 kg m^{-3}	*Not SI units, but more useful practically* kg m^{-3} = 0.001 g cm^{-3}
Soil cation exchange capacity (CEC)	Moles of ions per kilogram (dry soil)	mol kg^{-1}		*Ion should be specified*
Soil mineral N content	Gram (nitrogen) per square metre (ground) Kilogram (nitrogen) per hectare (ground)	g m^{-2} (g m^{-2} N) kg ha^{-1} (kg ha^{-1} N)	10^{-3} kg m^{-2} 10^{-4} kg m^{-2}	*Usually to 90 cm depth or 0–30 cm, 30–60 cm, 60–90 cm* 1 g m^{-2} = 10 kg ha^{-1} 1 kg ha^{-1} = 0.1 g m^{-2}

Table 4. Useful physical constants (Jerrard and McNeill, 1986)

Constant	Symbol	Value and unit symbol
Acceleration due to gravity	g	9.812 60 m s^{-2}
Avogadro's number	NA	6.022 045 $\times$ 10^{23} mol^{-1}
Boltzmann constant	k	1.380 662 $\times$ 10^{-23} J K^{-1}
Electron charge	e	1.602 1892 $\times$ 10^{-19} C
Electronvolt	eV	1.602 18 $\times$ 10^{-19} J
Planck's constant (per photon)	h	6.626 176 $\times$ 10^{-34} J s
Speed of light	c	2.997 924 58 $\times$ 10^8 m s^{-1}
Universal gas constant	R	8.314 41 J mol^{-1} K^{-1}

Table 5. Other data (Jerrard and McNeill, 1986; Farnworth, 1997)

Water

Density (293 K)	10^3 kg m^{-3} = 1 tonne m^{-3}
Specific heat (293 K)	4.19 J g^{-1} K^{-1}
Dynamic viscosity (293 K)	10^{-3} N s m^{-2}
Latent heat of vaporization (293 K, 101.3 kPa)	2454 J g^{-1}
Latent heat of fusion	334 J g^{-1}
Surface tension (293 K)	0.073 N m^{-1}

Air (dry)

Density (293 K, 101.3 kPa)	1.204 kg m^{-3}
Specific heat at constant pressure (293 K)	1.03 J g^{-1}
Thermal conductivity (293 K)	2.52 $\times$ 10^{-2} W m^{-1} K^{-1}
Dynamic viscosity (293 K)	1.82 $\times$ 10^{-5} N s m^{-2}

Area and volume calculations

Area of circle (radius r)	πr^2
Circumference of circle	$2\pi r$
Volume of cylinder (height h)	$\pi r^2 h$
Volume of sphere	$4/3\ \pi r^2$
Area of sphere	$(2r)^2 \times \pi$
Volume of cone	$(\pi r^2 h)/3$

References

Baron, D.N. (1994) *Units, Symbols and Abbreviations*, 5th edn. Royal Society of Medicine Press, London.

Czarnowski, M. (1996) Important measure units and symbols used in plant physiology. *Acta Physiologiae Plantarium* 18, 173–181.

Farnworth, J. (1997) *Agri Info: Guidelines for World Crop and Livestock Production.* John Wiley and Sons, Chichester, UK.

Jerrard, H.G. and McNeill, D.B. (1986) *A Dictionary of Scientific Units*, 5th edn. Chapman and Hall, London.

Monteith, J.L. (1984) Consistency and convenience in the choice of units for Agricultural science. *Experimental Agriculture* 20, 105–117.

Salisbury, F.B. (1996) *Units, Symbols and Terminology for Plant Physiology.* Oxford University Press, New York.

Index

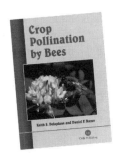
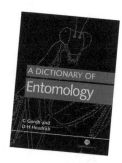
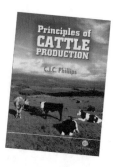